EMPLOYEE'S R

P9-BVC-244

I acknowledge receipt of Keller's *Employee's Guide to Food Safety*, which covers eight different food safety topics and a glossary of food safety-related terms. The topics covered in this handbook include:

Basic Microbiology & Foodborne Illness

Cross Contamination & Its Prevention

Foreign Material Detection

Hazard Analysis & Critical Control Points (HACCP)

Personal Hygiene Practices

Pest Control

Sanitation

Time & Temperature Controls

Glossary of Food Safety Related Terms

_____

Employee Name   (Please Print)

_____

Employee's Signature   Date

_____

Company

_____

Company Supervisor's Signature

NOTE: This receipt shall be read and signed by the employee. A responsible company supervisor shall countersign the receipt and place in the employee's training file.

Employee's Guide to Food Safety,

2nd Edition

©2000
J. J. Keller & Associates, Inc.
3003 W. Breezewood Lane, P. O. Box 368
Neenah, Wisconsin 54957-0368
Phone: (920) 722-2848
www.jjkeller.com

Library of Congress Catalog Card Number: 98-85795

ISBN 1-57943-863-6

Canadian Goods and Services Tax (GST) Number: R123-317687

Printed in the U.S.A.

2nd Edition, First Printing, June 2000

TABLE OF CONTENTS

INTRODUCTION TO THE EMPLOYEE

Your job is very important. As a person employed in the food industry, you are responsible for helping to make food products that many people across your region, the country, or even the world will eventually come into contact with. In addition, the food products you work with can carry potentially dangerous pathogens or other contaminants that can hurt or kill people when the contaminated food is eaten.

Because your job touches so many lives, and the food you work with is a potential carrier of such pathogens, you have to take special precautions when working. Your company has special hygiene, sanitation, and other cleanliness and food handling requirements that help to ensure that the food you help produce is safe for consumers to eat.

Part of your job is maintaining an awareness of these requirements at all times. It is very important that you know all the rules related to working with and around food at your company, and that you always follow them.

This handbook provides valuable food safety information that will enable you to keep yourself, the food you work with, and those who eat it safe. Your company may use this handbook in food safety training sessions. In addition, you can keep this handbook in your locker or designated storage area at all times, to serve as a handy reference for vital food safety information.

Maintaining constant awareness of food safety and following proper procedures means you contribute to a safe food product you can be proud to help produce.

BASIC MICROBIOLOGY & FOODBORNE ILLNESS

This chapter provides an overview of microbiology as it relates to foodborne illness and the many different types of harmful microorganisms that can contaminate food. Microbiology is the study of living things ordinarily too small to be seen with the naked eye and have to be viewed with the help of a microscope. These organisms are called microorganisms and are often called bacteria or "germs." Pathogenic bacteria are the types of microorganisms covered in this chapter.

Whether microorganisms are good or bad depends upon one's point of view. This is especially true in the food industry. Some types can be beneficial to humans. For example, some microorganisms are used to make vitamins, bread, cheese, yogurt, beer, and summer sausage. But others, like the ones covered in this chapter, can cause serious illnesses when contaminated food is eaten.

The food safety procedures followed at your company are designed to keep food safe from contamination of any kind. But microbial contamination is especially important, because bacteria that contaminate food can have a public health significance.

The term *public health significance* means there is a health threat to many members of the public if they eat contaminated food or come into contact with someone else who has eaten it and could pass disease on to them. For example, a physical hazard,

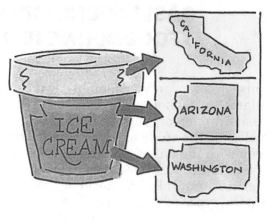

like a sliver of metal in a food product, may be swallowed and hurt only a single person, but a batch of ice cream contaminated with *Salmonella* bacteria can affect thousands of people across several states.

Your company takes several steps to prevent the contamination of food with potentially harmful bacteria. The main controls of bacteria are:

Time and temperature controls — Your company probably has some time and temperature controls in place to prevent growth of bacteria and molds in food. Keeping food for too long in the temperature Danger Zone (40-140 degrees Fahrenheit) can accelerate the growth of harmful bacteria. Proper time and temperature controls minimize growth.

Moisture — If a product is dry or has low water levels, then bacteria can't grow as easily.

Sanitation — If equipment is kept clean of food residue, it takes away bacteria's food source, preventing the microorganisms from growing.

Acidity — Bacteria do not like very acid conditions. We preserve food by adding acid, such as in pickling.

Restriction of travel — Preventing bacteria from moving from one place to another (cross contamination) controls their spread and growth. It is especially important to control cross contamination from raw to cooked product, and to use good personal hygiene to prevent spreading human bacteria, like *E. coli*.

In addition, your company may conduct laboratory tests on batches of food in an effort to detect bacterial contamination of outgoing food shipments. Unfortunately, despite the best efforts of food producers, some contaminated food inevitably slips into the public food supply and causes foodborne illnesses.

The bacteria that cause foodborne illnesses are called "pathogens" and are potentially deadly. Many of them are spread by the human hand; therefore, this chapter is dedicated to helping you understand their potentially deadly impact and the importance of following good personal hygiene practices and other food safety rules and procedures.

This chapter covers the following two main issues related to basic microbiology:

- Common pathogens and their prevention, and
- Workplace practices and procedures.

Common pathogens and their prevention

The bacteria that commonly cause foodborne illnesses have long, strange names, but they are not that complicated to understand. These pathogens either infect people directly when ingested, or produce toxins in the packaged product or person's intestines after the food is eaten. The results of foodborne illness could include diarrhea, stomach cramps, vomiting, headaches, fever, and even death.

The bacteria that cause foodborne illnesses are naturally present everywhere in our environment. This includes your workplace. Therefore, you need to be aware of these bacteria and how to prevent contaminating food with any of them.

Six types of pathogenic bacteria that you need to be aware of include the following:

- *Clostridium botulinum,*

- *Clostridium perfringens,*

- *Escherichia coli (E. coli),*

- *Listeria monocytogenes,*

- *Salmonella,* and

- *Staphylococcus aureus.*

This chapter will provide you with basic information on these six potentially dangerous bacteria. The information will help you understand how they are transmitted and the protective measures you must follow to protect the food you make from contamination with them.

Clostridium botulinum

Clostridium botulinum is a spore-forming bacteria typically found in soil, water, vegetables, grains, and the intestines of animals. Under the right conditions, the spores of this bacteria will grow and produce a toxin that is very poisonous. This toxin is one of the most deadly in the world. The disease botulism occurs when this potent toxin is eaten.

Botulism causes blurred or double vision, dry mouth, difficulty swallowing, and paralysis of respiratory muscles. Although it is not as common a disease as some of the others mentioned in this chapter, its fatality rate is high, and recovery for survivors takes months.

The threat of the spores growing and producing the toxin that causes botulism depends on a number of factors. For example, spores present on vegetables at the time of harvest can be difficult to remove completely by washing. When the contaminated vegetables are bottled or canned and steamed in a cooker that does not reach the necessary pressure and temperature, some *botulinum* spores may survive because they are highly resistant to heat. All the air once present in the containers is usually removed during the cooking process. Since *botulinum* spores only grow when there is no oxygen in a container, they can quickly produce the poisonous toxin when the containers of food are stored at room temperature.

Therefore, botulism poisoning occurs most often with improperly canned low-acid foods or with underprocessed home-prepared foods such as green beans. That's why time and temperature controls are very important and why the canning process is strictly regulated.

To control *Clostridium botulinum* special packaging may be used for some foods. For example, at one time this pathogen was a problem for vegetables in sealed plastic bags because once the vegetables used up all the oxygen, the spores grew and released the *botulinum* toxin. To prevent this from happening, some vegetables like mushrooms are now often packaged in special plastics or in containers with holes. Oxygen can then flow in and out of the package so the vegetables can "breathe." "Breathing" prevents the *botulinum* spores from growing and producing the deadly toxin.

Botulism poisoning can be prevented through proper methods of preserving and handling bottled or canned food. Proper heating, maintaining low pH and water activity, and using temperature controls during food processing are all essential to control this extremely dangerous pathogen. All employees must properly use the required food safety controls and work procedures at their facilities.

Clostridium perfringens

This bacteria, like *Clostridium botulinum*, also produces spores that grow only in little or no oxygen. It is found in soil and in the intestines of healthy people and animals like cattle, pigs, poultry, and fish. It is difficult to destroy this bacteria because its spores will grow very fast when conditions are right.

Clostridium perfringens food poisoning can occur when inadequately cooked meats and fish are eaten. This is often a result of food left for long periods of time in steam tables, buffets, or at room temperature. Symptoms of this illness include nausea and severe diarrhea. This illness is typically mild with symptoms usually lasting only a short while.

It's especially important that food is cooled quickly so the bacteria don't have time to grow. Keeping food out of the Danger Zone (40-140 degrees Fahrenheit) is very important. This is why shallow pans might be used to cool product faster. The same concept applies when reheating a food. Reheat food fast so bacteria don't have time to grow.

Using proper personal hygiene practices and time and temperature controls during cooling and heating of food will prevent the growth of this bacteria.

Escherichia coliform (E. coli)

E. coli is one of the most widely known and common bacteria. The reason *E. coli* is so common is that it is naturally found in the feces and intestines of people and animals.

E. coli can cause foodborne illness through raw milk, raw or rare ground beef, unpasteurized apple juice or cider, uncooked fruits and vegetables, and even person-to-person contact. However, transmission usually occurs by cross contamination of feces into food. This can be from animal feces at the slaughterhouse or on the dairy farm. But food contamination can also occur from human feces that are not thoroughly washed off your hands after you use the toilet.

E. coli causes gastroenteritis, with symptoms of diarrhea or bloody diarrhea, stomach cramps, fever, and nausea. Usually, it causes brief episodes of discomfort or pain for infected people, but occasionally, the diarrhea induces dehydration severe enough to cause even death. There's a rare strain of *E. coli* that causes a more severe illness with much worse symptoms, like kidney failure.

Good personal hygiene and sanitation practices must be followed to prevent contamination of food with *E. coli*. Washing and scrubbing your hands thoroughly after using the toilet is essential since this bacteria is commonly found in feces. Always follow your company's personal hygiene and sanitation practices to prevent foodborne illnesses.

Listeria monocytogenes

Listeria monocytogenes is an extremely hardy bacteria because it can survive in various conditions. It is resistant to cold, heat, salt, pH extremes, and even stomach bile. *Listeria* is found in milk, water, soil, and the intestines of healthy people and animals.

Contamination of food with this bacteria can happen in various ways. Vegetables can become contaminated with *Listeria* from the soil or manure used to fertilize the plant. Uncooked meat and dairy products can be contaminated with the bacteria when the animal is slaughtered or when milk is collected. *Listeria* can also be found in processed food that has been contaminated by employees.

Illness from eating food contaminated with *Listeria monocytogenes* is called listeriosis. Most cases of listeriosis are caused by ingesting contaminated milk, cheese, ice cream, some vegetables, and ready-to-eat meats such as hot dogs and luncheon meats.

Listeriosis has flu-like symptoms of sudden fever, chills, and sometimes nausea. If the infection is severe, symptoms such as intense headache, stiff neck, vomiting, delirium and coma can occur. Though not as common as other foodborne illnesses, death from listeriosis occurs approximately 25 percent of the time. Individuals most susceptible to this illness are pregnant women, small children, the elderly, and people with weakened immune systems, such as cancer patients.

Because of the severity of listeriosis, it is extremely important to control these bacteria through proper handling and control methods. *Listeria monocytogenes* is killed by sufficient cooking, pasteurization, or other heat treatments. However, unless good manufacturing practices are followed while handling processed food, contamination can occur. To reduce this risk, follow all personal hygiene and sanitation practices at your company.

Salmonella

Most of us have heard of *Salmonella*. This bacteria can be found in raw shell eggs and the intestines and feces of animals. *Salmonella* is most often present in contaminated raw eggs, milk, meat, poultry, dairy products, and seafood.

When food contaminated with *Salmonella* is eaten, it causes *salmonellosis* in humans. This disease is characterized by headaches, stomach pain, diarrhea, nausea, vomiting, and fever. It can even cause death through dehydration brought about by diarrhea and vomiting.

Correct processing controls, like sufficient heat treatment or the rapid development of acid in a fermentation process, are important controls to prevent *salmonellosis*. Avoiding cross contamination from raw to cooked food and maintaining correct storage temperatures are very important control measures as well.

Staphylococcus aureus (Staph)

Staph is not found in the same places as other types of bacteria already mentioned in this chapter. Rather, this bacteria is usually found in our noses, throats, infected cuts, and on our skin. Therefore, *Staph* can easily be transmitted to food by employees who practice poor personal hygiene habits or have uncovered sores or boils.

The disease this bacteria causes is called *staphylococcal* food poisoning. Like several other food poisoning bacteria, it's symptoms include nausea, cramps, vomiting, and diarrhea.

Staph produces a toxin that is heat resistant. If the bacteria in a raw ingredient has a chance to grow, the toxin produced may not be destroyed during a heat process, even though the bacteria itself will be killed. That food, containing the toxin, may still cause food poisoning. Poor refrigeration or inadequate heating methods speed up the growth of this bacteria. That's why it's very important to store ingredients at proper temperatures and to follow specific time and temperature controls while cooking food.

In addition, always practice good personal hygiene to prevent this bacteria from spreading. Wash your hands with sanitizing soaps or solutions provided by your company before returning to work. Do not touch your face, hair, earplugs, etc. while working directly with food.

Thoroughly cover any cuts, sores, or boils and let your supervisor know about them. Remember, an exposed wound may contain bacteria that could contaminate the products you work with. If you sneeze or cough directly on any food, let your supervisor know immediately. He or she will determine what to do with the product.

The table below summarizes the bacteria covered in this section, and the control methods you can use to prevent contamination of food with them.

Bacteria	Control Methods
Clostridium botulinum	Proper heat processing, maintaining low pH and water activity, cleaning methods, and temperature controls
Clostridium perfringens	Good personal hygiene practices, time and temperature controls, and rapid heating and cooling
Escherichia coli (E. coli)	Good personal hygiene and sanitation practices, thorough cooking, and proper refrigeration (time and temperature controls)

Bacteria	Control Methods
Listeria monocytogenes	Proper heat treatment, preventing cross contamination, and temperature controls
Salmonella	Good personal hygiene and sanitation practices, thorough cooking, proper refrigeration (time and temperature controls), and preventing cross contamination
Staphylococcus aureus	Good personal hygiene practices and time and temperature controls

Workplace practices and procedures

As you can see in the table above, good personal hygiene practices are the most common and best way you can prevent contamination of food with a foodborne pathogen. Along with cleaning and sanitizing, these practices are universal methods used to keep pathogens from contaminating food and causing foodborne illnesses. Thus, they are stressed throughout this handbook and in this chapter.

In fact, every chapter in this handbook touches on some aspect of controlling pathogens to prevent foodborne illness. You can do your part by following all of the work procedures and rules at your food company.

Chapter in The Handbook	Relation to Foodborne Pathogen Control
Cross Contamination	Keeps bacteria from getting into food
Foreign Material Detection	Detects anything that is not supposed to be in food
Hazard Analysis and Critical Control Points (HACCP)	Identifies what pathogens pose a threat to food and prevents them from affecting the food through various control methods

Chapter in The Handbook	Relation to Foodborne Pathogen Control
Personal Hygiene Practices	Prevents the spread of pathogens or other contaminants into food through good management practices that are always in place
Pest Control	Prevents pests from contaminating food (pests are common carriers of many pathogens)
Sanitation	Keeps areas free of the food or moisture that can allow bacteria to grow
Time & Temperature Controls	Controls the growth of bacteria in food

Due to the important relationship of cleaning and sanitizing, personal hygiene, and time and temperature controls to foodborne illness prevention, the most important recommended practices are repeated here.

Personal hygiene practices

By keeping yourself clean and following strict cleanliness rules at work, you keep the food you work with free from potential contamination of bacteria such as *Salmonella*, *Listeria*, and *E. coli*. Your personal hygiene can make the difference between a safe food product and one that can cause illness or even death to those who eat it.

Follow these good hygiene practices to ensure you keep the food you work with free from contamination with foodborne pathogens:

- Maintain adequate personal cleanliness by bathing every day and wearing clean clothes to work;

- Wash hands thoroughly before beginning work, every time you use the restroom, and whenever your hands may have become soiled or contaminated;

- Remove all jewelry and other objects that might fall into food when you are working;

- Wear and keep gloves, hair restraints, smocks, plastic sleeves, or other personal protective equipment (PPE) in good condition;

- Store clothes and other personal belongings in designated areas separate from food production and storage areas or utensil/equipment washing areas; and

- Keep beverages, food, candy, chewing gum, and tobacco separate from food production and storage areas or utensil/equipment washing areas.

Time & temperature controls

Time and temperature controls are an extremely effective means of controlling the growth of bacteria.

Bacterial growth can be slowed and even stopped by sufficient heating and cooling methods. This is why some food facilities are cooled throughout and why you may have time limits to work on some types of food.

Follow these rules to ensure the correct control of time and temperature for a safe food product:

- Always follow established work procedures and practices

for the product you are working on.

- Don't prop open doors to a refrigerator, freezer, or any other cooled work areas.

- Never alter the specified time to heat, blanch, or cook a product.

- If in doubt as to whether a food had reached the correct processing temperature, use a thermometer to check.

- When timing of procedures or practices is necessary, pay careful attention to the clock to be sure to meet the correct time requirement.

- If you think a time or temperature control may have been violated, tell your supervisor or quality control person. He or she can test the batch or make a decision as to what to do.

In conclusion

Bacteria are constantly changing and adapting to their environment. We now have *Salmonella* that are resistant to antibiotics, *E. coli* that survive acid conditions, and other bacteria that are becoming increasingly heat resistant. So we have to be constantly vigilant to follow the rules and do everything correctly so these bacteria don't contaminate the food our company produces and makes someone sick.

NOTES

NOTES

BASIC MICROBIOLOGY & FOODBORNE ILLNESS REVIEW

1. Microorganisms are:
    a. Living organisms so small they can only be seen with the help of a microscope
    b. Living organisms large enough to be seen with the naked eye
    c. Organisms that are made of "micros"
    d. None of the above

2. Microorganisms are found in:
    a. Food
    b. Soil
    c. Intestines of people and animals
    d. All of the above

3. Microorganisms are used to make:
    a. Bread
    b. Cheese
    c. Summer sausage
    d. All of the above

4. Which is not a harmful bacteria:
    a. Listeria
    b. E. coli
    c. Zoology
    d. Salmonella

5. Disease caused by harmful bacteria is called:
    a. Microbial poisoning
    b. Foodborne illness
    c. Food contamination
    d. Food infection

6. A common symptom of many foodborne illnesses include:
    a. Vomiting
    b. Blindness
    c. Sleepiness
    d. None of the above

7. A procedure food companies use to try to control bacteria is:
   a. Refrigeration
   b. Cleaning and sanitizing equipment
   c. Cooking
   d. All of the above

8. The most important personal hygiene practice to control the spread of bacteria is:
   a. Wearing hair restraints
   b. Washing your hands
   c. Bathing daily
   d. None of the above

9. What time and temperature practices are recommended to control bacteria:
   a. If in doubt as to whether a food had reached the correct processing temperature, use a thermometer to check
   b. Don't prop open doors to a refrigerator, freezer, or any other cooled work areas
   c. Never alter the specified time to heat, blanch, or cook a product
   d. All of the above

10. You must always follow food safety rules and procedures because bacteria can:
    a. Change and adapt to their environment
    b. Cause foodborne illness
    c. Contaminate food
    d. All of the above

CROSS CONTAMINATION & ITS PREVENTION

Preventing cross contamination is one of the basic goals of food safety regulations. Cross contamination is the contamination of food with other materials. It's called cross contamination because its contamination across boundaries, something that is moved from one thing to another. Cross contaminants include anything in a food product that should not be there. This includes hazards like wood, glass, metal, chemicals such as pesticides or sanitizers, allergenic foods like peanuts, and potentially harmful bacteria like *E. coli*.

This chapter covers several basic concepts that will help you understand and prevent cross contamination of food. These topics include:

- Relation of cross contamination to other chapters in this handbook,

- Relation of cross contamination to food safety regulations,

- Chemical cross contamination,

- Bacterial cross contamination,

- Allergens,

- Keeping potential contaminants separate from food, and

- Practices to detect and eliminate cross contamination.

Relation of cross contamination to other chapters in this handbook

Cross contamination is closely related to many of the other topics covered in this handbook because cross contamination prevention is the common goal of the other subjects covered in this handbook. See the table below for the relationship each of the chapters has to preventing cross contamination. The italicized words indicate the specific relationship to cross contamination prevention.

Chapter	Relation to Cross Contamination Prevention
Basic Microbiology & Foodborne Illness	Many allergens and pathogens are from sources external to the food itself and *must be kept from contaminating food.*
Foreign Material Detection	Foreign material detection is the *detection of cross contamination that has occurred.*
Hazard Analysis and Critical Control Points (HACCP)	HACCP requires the identification of physical, chemical, and microbiological hazards that pose a threat to food, *and methods to prevent them from cross contaminating food*
Personal Hygiene Practices	Personal hygiene practices focus on *preventing cross contamination of food by humans* from personal items like jewelry, illness, hair and skin, etc.
Pest Control	Pest control is the *prevention of cross contamination of food by pests* such as insects, rodents, birds, etc.
Sanitation	Sanitation includes all of the practices and *procedures used to keep a food production or processing facility clean and the food it produces free of contaminants, some of which are cross contaminants* and potentially harmful bacteria.
Time & Temperature Controls	Time and temperature controls stop or *minimize the growth of bacteria that can contaminate food.*

Relation of cross contamination to food safety regulations

Preventing cross contamination is so central to food safety that it is at the heart of the food safety laws that regulate our food industry. It is the main goal behind food safety regulation.

The first official food law ever made, the Food & Drug Act passed in 1906, was written to prevent cross contamination of food. It stated simply that a food must not be adulterated. FDA's Current Good Manufacturing Practices (CGMPs) still have the same objective — to prevent adulteration of food that is produced for human consumption.

Adulterated food is defined by the CGMPs as food that "is manufactured under such conditions that it is unfit for food" or food that "has been prepared, packed, or held under insanitary conditions whereby it may have become contaminated with filth." Prevention of cross contamination is clearly the primary goal of CGMPs.

In addition, the main goal of Hazard Analysis and Critical Control Points (HACCP) is to prevent cross contamination. HACCP requires the identification of any physical, chemical, and microbiological hazards associated with each product made by conducting a *hazard analysis*. The *critical control points* part of HACCP establishes the methods to prevent these hazards from contaminating food.

It's very important to obey the food safety rules in your plant to prevent cross contamination.

Chemical cross contamination

Chemical cross contamination has its own section in this chapter because of the particularly hazardous nature of this type of contamination of food. In addition, chemical cross contamination is not covered as deeply as other types of contamination in other chapters of this handbook.

A wide variety of chemicals are routinely used in food production facilities to keep them clean. Many of them, if used improperly, can be toxic (poisonous) or otherwise dangerous to humans. The table below gives you an idea of some of the kinds of chemicals in use in food production facilities.

Chemicals In Food Facilities	Use
Pesticides	Pest Control
Lubricants and paints	Plant and equipment maintenance
Food additives and processing aids	Production of food
Cleaners and sanitizers	Sanitation

Many of these chemicals must be used in close proximity to or in actual food production processes or areas. At the same time, they must be kept from contaminating food.

When properly used, these chemicals are both safe and necessary to the food production process. However, when used improperly or allowed to contaminate food, some chemicals are capable of causing illness or even death.

The following are recommended practices for working with chemicals in food production areas to avoid chemical cross contamination of food. Follow these guidelines and all relevant work rules when working with or around chemicals in your workplace:

- Always put food and packaging materials away or make sure they are properly covered before cleaning or applying pesticides.

- Use only approved chemicals specified by your company for the procedure or practice you are doing.

- Use each chemical according to the manufacturers recommended guidelines or rules printed on the label.

- Don't use chemicals if you have not been properly trained. Talk to a supervisor for more information if you're asked to use a chemical whose hazards you are unaware of.

- Store chemicals in their designated spot when not in use.

- Always make sure the proper label is affixed to the container. If you find an unlabeled container, relabel it only if you know its contents. Otherwise, bring it to your supervisor, who will determine the contents of the container or dispose of it properly.

- Make sure food production work areas are cleaned, rinsed, and free of chemical residue before resuming food production after using cleaning, sanitizing, maintenance, pesticide, or other potentially contaminating chemicals.

- Know where the Material Safety Data Sheets (MSDS) are kept in your work area(s).

- Don't over-lubricate equipment because excess oil could contaminate food.

- Be sure that a thorough clean up is done after work on a piece of equipment is complete to ensure that nuts, bolts, or metal filings are not left in a place where they could potentially contaminate food.

Bacterial cross contamination

Bacteria do not have legs, so they cannot move around on their own. They have to hitch a ride on people, equipment, packaging materials, etc. to move from one place to another. This transfer of bacteria is called cross contamination. Good personal hygiene practices and obeying all the CGMP rules in the plant, like handwashing and using footbaths, are extremely important in preventing bacterial cross contamination. Also, never taste the food you're working with. Bacteria from your mouth could also contaminate the food and make people sick.

Another good control to help prevent bacterial cross contamination is to separate raw product from cooked product. A raw product can contain various types of bacteria. If the product is cooked properly, the bacteria will be killed. Therefore, it's very important cooked food does not get contaminated after the heat process because the food may not be cooked again before it is eaten.

Separating raw and cooked product might mean your company has separate raw and finished work areas and/or separate equipment in raw and finished product areas of the facility. The equipment may even be color-coded to help employees identify which equipment should be used in which area. This prevents accidental contamination.

Allergens

Some consumers are very sensitive to certain foods. This is known as a food allergy. If a sensitive person eats or has contact with a food containing a known allergen, it could trigger a severe and potentially deadly reaction in their body. Sufferers of food allergies can exhibit a variety of symptoms after eating an allergy-causing food like a tingling sensation in the mouth, swelling of the tongue and throat, difficulty breathing, hives, vomiting, stomach cramps, diarrhea, and loss of consciousness.

A food allergy is a very serious condition because there is no cure or medication to prevent an allergic reaction from occurring. Strict avoidance of the allergy-causing food is the only way to avoid a reaction. The scary part is that sometimes it only takes a tiny amount of the allergen to cause the reaction. The most common food allergens include:

- Peanuts,

- Tree nuts (almonds, walnuts, pecans, and cashews),

- Shellfish (shrimp, crab, and lobster),

- Fish,

- Eggs,

- Wheat,

- Milk, and

- Soy.

People who suffer from a food allergy always read the label of a food product to make sure there are no allergens they're sensitive to in that food. Food companies have to ensure all foods are properly labeled and that no ingredient not declared on the label mistakenly ends up in the product.

Because of the serious nature of food allergens, it takes the awareness and contribution of everyone working at your company to avoid this type of cross contamination. Some of the controls you and your company can put in place and follow are:

1. Always check the ingredients before you add them to a batch. Check each ingredient against the formula or recipe sheet and mark them off as you add them. Read the labels of ingredients carefully. Things like vegetable oil could easily be mistaken for peanut oil.

2. If possible, use separate equipment only for products that contain a known allergen. This prevents residue that may be left on equipment from contaminating other foods. If that is not possible, the equipment must be thoroughly cleaned after production of a food containing an allergen before a non-allergen containing food is made.

3. Try to schedule allergen-containing products at the end of the day or week so that there is time to clean the equipment before the next production run.

4. Control rework. If you use rework in your product, you have to make sure it is compatible with the product you are making. If the rework has egg in it, it can only be used in a formula that also has egg in it and for which egg is declared on the label.

5. Be careful to use the correct packaging. Mistakes have been made when a chocolate chip cookie with nuts gets into a bag of chocolate chip cookies without nuts in them. The two cookies may look the same, so a consumer wouldn't be aware of the hidden danger.

6. If you work in receiving, make sure the correct ingredients and raw materials are being delivered. Check to ensure that what is coming off the truck is the same as what's on the order form.

7. Like bacterial cross contamination, if you have nuts in your plant, there might be restricted areas for people and restricted color-coded equipment that can only be used for those products.

8. In storage or preweight areas, it is important not to mix ingredients that might not be compatible. Use separate scoops for each ingredient and clean out bins between uses.

Keep potential contaminants separate from food

Keeping potential contaminants separate from food is the easiest and most effective way to prevent cross contamination. If a potential contaminant is not in a food production area, it cannot possibly get into the food being produced. Following are examples of separation:

- Store personal effects like purses or magazines, non-work clothes, food, and beverages separate from all food work areas. This is required under CGMPs. Your company informs you of the designated places to store these personal items. When stored separately, these items have no opportunity to contaminate food.

- Always store chemicals such as pesticides, cleaning chemicals, and equipment maintenance lubricants and oils in designated areas separate from food production areas. There they will not spill, leak, or be accidentally added to the food being produced.

- Keep pests out of the facility through whatever measures you can within your company's pest control program. If pests are not in the food areas, they cannot contaminate food.

- Inspect incoming materials and raw ingredients as necessary to detect any contaminants before the ingredient is introduced into the facility or food production process. Inspection takes place at initial entry or other separate areas, before the potentially contaminated raw ingredient has a chance to contaminate food product.

It's important to keep as many non-food and non-work-related items separate from food production areas, to prevent contamination of food ingredients and finished food product.

Practices to detect and eliminate cross contamination

Sometimes cross contamination does occur either when rules are disobeyed, control measures are inadequate, or through human error or accident. That's why we cannot rely solely on the prevention of cross contamination.

We must also have practices in place to check food during and after production, to detect cross contamination that may have occurred. Some common detection practices include:

- Visual inspections of food,

- Use of metal detectors, magnets, x-rays, or other detection equipment,

- Use of screens or sifters, and

- Microbiological testing.

These and other contamination detection practices are covered in the *Foreign Materials Detection* chapter, which is dedicated solely to the topic of detecting food contaminants. See this chapter for further information on cross contamination detection techniques.

In summary

Because cross contamination covers so many areas, use the information found throughout this handbook to prevent cross contamination of food. Preventing cross contamination means careful work conducted according to established procedures. Follow the food safety rules at your company to prevent cross contamination of food.

NOTES

NOTES

CROSS CONTAMINATION

Name _____

Date _____

CROSS CONTAMINATION REVIEW

1. Cross contamination is:
   a. Contamination of food with crosses
   b. Contamination of food with other materials
   c. Contamination of food with ingredients
   d. None of the above

2. Which is not a possible contaminant:
   a. Pesticides
   b. Wood splinters
   c. Bacteria
   d. Moon beams

3. Chemical contaminants include:
   a. Water
   b. Cleaning chemicals
   c. Milk
   d. All of the above

4. To prevent chemical cross contamination always:
   a. Store chemicals in their designated spot when not in use
   b. Make sure food production areas are cleaned, rinsed, and free of chemical residue after sanitation
   c. Put food and packaging materials away or cover them before cleaning or applying pesticides
   d. All of the above

5. Bacteria can cross contaminate food by hitching a ride on:
   a. People
   b. Equipment
   c. Packaging materials
   d. All of the above

6. To prevent bacterial cross contamination always:
   a. Wash your hands
   b. Use footbaths and hand dip stations
   c. Separate raw and cooked products
   d. All of the above

7. Physical hazards include:
   a. Metal
   b. Dust
   c. People
   d. Water

8. Food allergens include:
   a. Peanuts
   b. Milk
   c. Shellfish
   d. All of the above

9. Which is not used to detect cross contamination in food:
   a. Metal detector
   b. Screwdriver
   c. Visual inspection
   d. Screens and sifters

10. Cross contamination occurs when:
    a. Rules are disobeyed
    b. Control measures are inadequate
    c. There is human error or accident
    d. All of the above

FOREIGN MATERIAL DETECTION

Foreign material detection is a way your company tries to locate objects that might be in food, but that should not be there. It is a process your company must perform on all food at several stages during the production process.

Foreign materials in food are also commonly referred to as "physical hazards" or "contaminants." All three terms are used in this handbook to refer to physical objects that should not be in food, but that sometimes end up there.

Foreign material detection makes good business sense, because consumers want food free of contamination with foreign materials.

In addition, foreign material monitoring is required by several government standards, including:

- **Hazard Analysis and Critical Control Points (HACCP)** — A regulation that requires the identification and evaluation of physical, biological, and chemical hazards.

- **Food and Drug Administration's (FDA) Current Good Manufacturing Practices (CGMPs)** — A regulation that specifies effective measures must be taken to protect food against foreign material contamination.

That is *why* companies must detect foreign materials in food. Based on these requirements, your company must have processes and procedures in place to detect and prevent foreign materials in food.

The rest of this chapter will familiarize you with the *when, what, where,* and *how* of foreign material detection, as well as control measures to avoid foreign material contamination of food. Your company will tell you *who* conducts foreign material detection. It may very well be you.

What to detect and where it comes from

There are a wide variety of foreign materials that must be eliminated in the food production process. These include items that may come into the food production plant in raw materials, may be picked up in the plant, or may be brought into the plant by employees.

The table below lists some common foreign materials and their source or cause that could end up in food.

Foreign Material (physical hazard)	Source or Cause
Metal	Nuts, bolts, screws, grinders, mixers, knives, scoops, shovels, meat hooks, bb shot, nails from pallets
Glass	Light bulbs, watch crystals, thermometers, bottles
Wood splinters	Crates, pallets, equipment bracing, overhead structures, brooms
Insects	Environment, incoming ingredients
Nut shells or fruit pits	Ingredients/raw materials
Plastic	Box or combo bands, tank covers, packaging materials, ingredient bags
Hair, gum, jewelry, pens, cigarette butts, etc.	Poor personal hygiene practices by employees

As you can see, there are a wide variety of foreign materials, and an equally wide variety of sources for them. None of these items are supposed to be in food, most of them are potentially harmful, and all of them are unacceptable.

The topic of insects, a common foreign material found in food, is discussed in a separate chapter of this handbook. See the *Pest Control* chapter in this handbook for more information on insect detection and elimination, as well as for information on rodents and birds.

When to conduct foreign material detection

The most important times to conduct foreign material detection is when stock is coming into the plant and when finished food product is about to be packaged or is about to leave the plant. However, attention must be paid to foreign material detection and prevention at all points during the food production process, not just the beginning and the end.

Incoming materials

Incoming materials can harbor pests, dirt, and debris. The pallets and crates that they are carried on or in are a frequent source of wood splinters, nails, and other physical contaminants. The CGMPs specifically require incoming raw materials to be inspected for potential contaminants.

Outgoing food

Food on its way out of the plant is an even more important inspection point. This is the last chance to detect foreign materials before the food leaves the plant and is distributed for consumption. In addition, since all stages of the food production process are complete by this point, outgoing food inspection is done after the last opportunity for foreign materials to contaminate the food.

Ongoing detection

Ongoing inspection and detection is also important because there are many points during the production process when foreign materials could get into food.

How To Detect

Using in-line equipment for detecting and removing foreign materials is a common method of handling incoming raw materials. Similar in-line equipment, such as metal detectors, magnets, and occasionally x-rays, may be used during the production process or on the finished product  before it is shipped. If equipment such as this is used, the operator must be trained on its proper use, and the company must do proper installation, regular maintenance, and calibration.

Another inspection method of incoming and outgoing product, perhaps the most common, is visual inspection. That means an employee, like you, looks at the food to spot foreign materials with the naked eye.

This can be a highly effective technique if used properly. If you are expected to visually identify foreign materials in food, you will be trained on what to look for and at what stage in the production process to look for it. Examples of inspection situations include:

- Inspection tables or belts on a line to examine raw fruits and vegetables before they are processed. If you do this type of inspection you will be trained to take out stones, stems, leaves, wood, etc.

- Inspection of incoming raw materials such as beef, chicken, or pork. This inspection would occur near receiving as the meat is placed on a production line. You would be trained to look for foreign materials such as cardboard, wood, dirt, and metal objects such as meat hooks, shovels, and thermometers.

Finally, there are several pieces of equipment designed for specific foreign material detection or removal functions. A list of the most common ones and their functions are described in the following table:

Detection/Removal Equipment	Function
Magnet	Removes hazardous metals
Metal detector	Detects various sizes of metal pieces
X-ray machine	Detects any material in food that has a different weight from the food
Screen, filter, or sifter	Removes foreign objects larger than the size of the equipment openings
Aspirator	Removes material lighter than product
Riffle board	Removes stones from dry beans and field peas
Bone separator	Removes bone chips from meat and poultry products
Inversion equipment for canning or bottling lines	Mechanically inverts each bottle or can so anything inside falls out

It is likely that at least one of the types of equipment listed in this table is used at your company.

Control measures

It is often said that the best offense is a good defense. That's why the most important way to protect against foreign material contamination of food is to have control measures in place to prevent it from happening in the first place.

There are several common control measures that can prevent foreign materials from getting into food products. The best control measures you and your company can use to protect food from foreign material contamination are listed here. Always:

- Practice good personal hygiene. This keeps the easily preventable foreign materials like jewelry, pens, gum, and other personal items from getting into food. See the *Personal Hygiene* chapter for detailed information about good personal hygiene habits.

- Inspect incoming raw ingredients *before* they are put into the food production process. Clean those ingredients that are cleanable (like fruit) and sift the siftable (like grains and flours).

- Take care as you add raw materials or ingredients into a product. For example, when opening and dumping bags, be careful to keep pieces of paper or plastic from getting mixed in with the ingredients.

- Have quality and inspection agreements with suppliers, so they do the work of preventing contamination in raw ingredients before they're sent to your company.

- Maintain constant awareness of the production environment and watch for signs of potential contamination sources like flaking paint, rusty equipment, and loose screws, nuts, or bolts.

- Practice good housekeeping at all times to keep work areas clean and free of debris. Maintenance staff especially needs to maintain awareness of good housekeeping, cleaning up loose objects or metal filings once they have completed their work.

In conclusion

As you can see, employee practices are an essential part of control measures. Do your part to control foreign materials in the food your company produces.

Having control measures in place doesn't mean your company doesn't have to use foreign material detection devices and procedures. It does mean that when they do use those devices and procedures, there will be fewer foreign materials to detect.

FOREIGN MATERIAL

NOTES

NOTES

Name _____

Date _____

FOREIGN MATERIAL DETECTION REVIEW

1. Foreign material detection is:
   a. Finding objects in your morning coffee before going to work
   b. Locating objects in food that should not be there
   c. Sanitizing of equipment and food work areas
   d. None of the above

2. Which terms are used to mean the same thing as foreign material:
   a. Physical hazard
   b. Contaminant
   c. Both of the above
   d. Neither (a) nor (b)

3. Foreign materials can come from:
   a. Raw materials
   b. Inside the food facility
   c. Employees
   d. All of the above

4. Types of foreign materials include:
   a. Metal
   b. Glass
   c. Insects
   d. All of the above

5. Poor personal hygiene practices are the cause of which types of contaminants:
   a. Chewing gum
   b. Human hair
   c. Cigarette butts
   d. All of the above

6. It is important to conduct foreign material detection:
   a. When raw materials are received and when food is about to leave the plant
   b. During the cooking process
   c. After the food has arrived at the food store
   d. All of the above

7. Incoming raw materials can contain which foreign materials:
    a. Pests
    b. Dirt
    c. Debris
    d. All of the above

8. Which equipment is used to detect foreign materials:
    a. A thermometer
    b. Magnets and metal detectors
    c. A refrigerator
    d. None of the above

9. Visual inspection means:
    a. Looking through a magnifying glass
    b. Using magnets and metal detectors
    c. Looking for foreign materials with the naked eye
    d. None of the above

10. Which control measures help prevent foreign materials from getting into food:
    a. Good personal hygiene practices
    b. Inspection of raw materials before adding them to food
    c. Good housekeeping measures
    d. All of the above

HACCP

HACCP is an acronym you may or may not have heard of before. It stands for **H**azard **A**nalysis and **C**ritical **C**ontrol **P**oints. HACCP is a logical, simple, effective, but highly structured system of food safety control. In the United States, the Food and Drug Administration (FDA) and the United States Department of Agriculture (USDA) have made HACCP a rule that many food companies must follow.

Your supervisor or quality assurance personnel can tell you whether your company must comply with HACCP, or whether it chooses to voluntarily use the HACCP system. If your company is required to or chooses to follow HACCP, then this chapter will provide you with valuable background information, so you can do your part to keep your company in compliance with HACCP requirements.

This chapter will give you an overview of HACCP and explain its relation to record-keeping, inspections, and verifications that you may be called upon to do in your job.

History of HACCP

HACCP is not new. It was first developed in the 1960's as a spin-off of the United States space program. The government wanted to make sure that the food eaten by astronauts would not harm or make them sick in space. They wanted food that was completely free of any bacterial pathogens, toxins, chemicals, and physical hazards that could possibly cause illness or injury to the astronauts.

As a result, HACCP was soon developed as a systematic approach to eliminate these potential hazards to produce the safest food possible.

Overview of HACCP

HACCP is based on a common-sense application of technical and scientific principles to the food production process from beginning to end. HACCP identifies the biological, chemical, and physical hazards that are reasonably likely to cause illness or injury to a consumer for every food product made at your company. Preventive measures are then put into place to stop these hazards before they begin.

The principles of HACCP apply to all phases of food production. This includes basic agriculture, food preparation and handling, food processing, foodservice, distribution systems, and even consumer handling and use. However, this chapter focuses mainly on the application of HACCP to food preparation, handling, and processing such as your company does.

HACCP is not a stand-alone food safety system. It is just one building block of a complete food safety program. Good sanitation, personal hygiene, microbiology, and pest control programs are all necessary to safe food production. These are sometimes referred to as prerequisite programs because they need to be done before a HACCP system can be effectively implemented. For example, you have to start with clean equipment if you expect to produce clean food. Your company probably provides food safety training on these and other food safety topics covered in this handbook.

HACCP requires that everyone involved in food production have enough information about the food and its production process to be able to identify where and how food safety problems could occur. A HACCP program deals with controlling factors that affect the ingredients, product, and process in order to make food safely. The HA (Hazard Analysis) part involves thinking of all the possible things that could go wrong with an ingredient or process that could cause harm to a consumer. The CCP (Critical Control Point) part involves identifying where in the process those potential hazards can be controlled.

Seven principles of HACCP

HACCP is a systematic approach to food safety that consists of seven basic principles. They are used to develop a plan that prevents the contamination of food. These principles include doing each of the following step by step:

1. Hazard Analysis — Prepares a list of steps in the process where significant biological, chemical, and physical hazards occur and develops preventive measures used to prevent, eliminate, or reduce each hazard to an acceptable level of safety.

2. Critical Control Points (CCPs) — Identifies a point, step, or procedure in the process at which a food safety hazard can be controlled.

3. Critical Limits — Develops maximum and/or minimum values of safety for a hazard at a CCP to prevent, eliminate, or reduce the hazard to an acceptable level of safety. If a product is outside the critical limit, it is considered potentially unsafe.

4. Monitoring — Conduct observations or measurements in the process to determine whether a CCP is under control.

5. Corrective Actions — Establishes and implements a procedure to be followed when monitoring indicates that there is a deviation from an established critical limit. A corrective action tells the operator what to do if something goes wrong.

6. Recordkeeping — Develops and maintains records that document the HACCP system. The company keeps records so it can prove the product was produced in a safe manner.

7. Verification — Establishes and implements activities to verify the adequacy of the HACCP plan and that the system is working correctly.

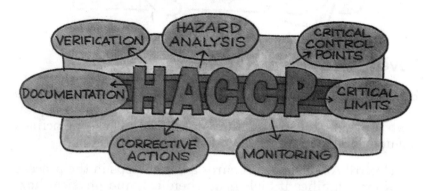

As a food production employee, it's possible you don't work directly with any of the seven basic principles of HACCP, but odds are that you do. Activities like monitoring, implementing corrective actions, and keeping records for documentation are all HACCP related activities you might be involved in at some time or another in your job.

Practical implementation of HACCP

In the development of a HACCP plan, there are several preliminary tasks that need to be accomplished before the seven HACCP principles can be utilized. First, a HACCP team must be assembled. A HACCP team is a group of people who are responsible for developing, implementing, and maintaining the HACCP system. The team consists of people who have specific knowledge and expertise of the food products and processes at your company.

A HACCP team typically includes people from a variety of areas within the company such as engineering, production, sanitation, maintenance, microbiology, and quality assurance/control. You might even participate on a HACCP team related to the food product or process you work with.

The HACCP team has several responsibilities. First, the team makes a list of all the food products made at your company. A separate HACCP plan must be developed for each product or group of products that is processed at your company.

Once a product or group of products needing a HACCP plan is identified, a general description must be written for each. The description can include the product name, packaging type, length of shelf life, labeling instructions, method of storage and distribution, and intended use and consumer. This information helps assist the team while they're developing the HACCP plan.

Finally, the last step the team takes prior to implementation of the seven principles of HACCP is to develop a flow diagram for each product described. The purpose of the flow diagram is to provide a clear, simple picture of the steps involved in the food production process from beginning to end. Once the flow diagram is complete, the team is ready to begin developing the HACCP plan.

The seven principles of HACCP in practice

After the HACCP team has completed the preliminary tasks discussed above, they are ready to apply the seven principles of HACCP to each product they've identified and described. These seven principles provide the bulk of the written HACCP plan and documentation.

Hazard Analysis

First, the team must conduct a hazard analysis of the food. This analysis identifies all significant chemical, physical, and biological hazards for each step identified in the flow diagram drawn for each product. HACCP covers all types of potential food safety hazards, whether they are naturally occurring in the food, contributed by the environment, or generated by manufacturing mistakes.

Consumers fear all hazard types. Chemical hazards such as poisons and allergens can cause severe illness or even death. Physical hazards like metal, glass, or wood are commonly found by consumers in food and result in customer complaints. 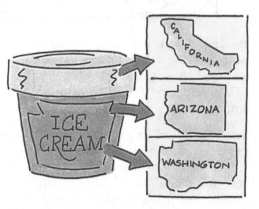 Biological hazards are the most serious type from a public health perspective because they can effect a great deal of people. For example, a physical hazard such as a piece of metal in a food product is likely to be swallowed by only one person eating that food, but a batch of ice cream contaminated with *Salmonella* can affect thousands of people across several states.

The table below includes examples of common biological, chemical, and physical hazards the team may identify during its analysis. Many other similar hazards exist.

Biological hazards	Chemical hazards	Physical hazards
bacteria	cleaning supplies	metal
viruses	pesticides	glass
molds	machinery lubricants	plastic
parasites	allergens	wood

When identifying all the food safety hazards for each product, the team needs to consider several things. This includes examining the chemical, physical, and microbiological characteristics of the food product, how its processing affects those characteristics, and interactions between ingredients. Each food company is unique in how it makes a product, so the team must consider how the food is made at your particular plant.

After the HACCP team has made a list of all the potential hazards for each product, they must decide which of these hazards must be addressed in the HACCP plan. During this stage, the severity and likely occurrence of each potential hazard is closely evaluated.

The next step is to develop preventive measures. Preventive measures are physical, chemical, or other methods that can be used to control each hazard before they become a safety issue. For example, having employees physically inspect incoming raw materials for hazards such as wood, metal, or glass could be a preventive measure used to eliminate these hazards from further contaminating food and potentially harming a consumer. Other preventive measures used to control hazards could include time and temperature controls to kill bacteria or washing equipment between production runs to eliminate cross contamination. You may perform a similar preventive measure procedure at your company.

The steps in the process where these preventive measures are implemented are called control points.

Critical Control Points (CCPs)

After identifying each hazard and developing preventive measures, the next step is to identify critical control points. A Critical Control Point (CCP) is any point, step, or procedure in the process at which control can be applied and, as a result, a food safety hazard can be prevented, eliminated, or reduced to an acceptable level of safety.

The following are examples of some common CCPs:

- Chilling of food when appropriate;

- Cooking to a specific time and temperature to destroy harmful pathogens;

- Product formulation controls (such as adding culture or adjusting pH and water activity);

- Certain processing procedures (such as filling and sealing cans);

- Metal detection; and

- Certain slaughter procedures.

Critical limits

After the CCPs have been identified, the team must establish critical limits for each CCP. A critical limit is a maximum and/or minimum value of safety. A critical limit is used to distinguish between safe and unsafe operating conditions at a CCP. Therefore, if a product is outside the critical limit, it is considered potentially unsafe.

Critical limits may be based upon factors such as temperature, time, physical dimensions of a product, humidity, moisture level, pH, etc. In some cases, federal regulations establish the critical limit (e.g., cooked poultry's critical limit is 160 degrees Fahrenheit under USDA regulations). If there is no regulatory guidance, the HACCP team can consult scientific literature, experimental studies, and experts in the field to determine appropriate critical limits.

Monitoring

Monitoring of critical limits at CCPs is the next principle. Monitoring is observations or measurements to assess whether a CCP is under control. These observations or measurements are recorded on a form and kept for future use. Monitoring is essential to a HACCP system. It warns if there is a trend towards loss of control, so the company can take action to bring a process back into control before a critical limit is violated.

Monitoring procedures and their frequency are determined by the HACCP team and recorded in the HACCP plan. But the monitoring itself is the first activity of the seven HACCP principles that may be done by someone other than a HACCP team member. Anyone at the company trained in the proper procedures may perform monitoring at a CCP. Monitoring may be continuous (such as metal detection) or non-continuous (such as visual examinations or measuring of pH, water activity, and product temperatures).

If your job involves monitoring, whether HACCP-related or not, always follow the established procedures you were instructed to use. Monitoring procedures must give an answer right away, so corrective action can be taken immediately if something is wrong.

If you ever suspect a problem with the results from your monitoring, or are unsure how to interpret them, ask a supervisor or quality assurance person for help. In addition, follow all company policies and practices with respect to monitoring, including using specified equipment and personal protective equipment (PPE).

Corrective actions

Corrective actions are simply the procedures to be followed when failure to meet a critical limit is discovered during the monitoring process. The purpose of corrective actions is to prevent foods that may be hazardous from reaching consumers.

Corrective actions will be determined and documented in the HACCP plan by the HACCP team. Corrective actions must define exactly what has to be done if something goes wrong. This includes indicating the responsibilities of each person who may be involved. Employees trained in monitoring may also be trained to perform the appropriate corrective actions developed by the HACCP team.

Some examples of corrective actions may include the following:

- Immediately adjust the process and hold product for further evaluation;

- Empower employees to stop the line when a deviation occurs and hold all product until a HACCP-knowledgeable person has assessed the situation and determined appropriate action; and

- Rely on an approved alternate process that can be substituted for the deviating one.

Regardless of what corrective action is taken, records must be kept identifying the specific deviation, what corrective action was taken, details of the incident, and actions planned to prevent a reoccurrence. A corrective action must do two things: first, it must bring the process back in control, and second, it must tell the operator what to do with the product, for example, put it on hold.

Recordkeeping

It is evident that HACCP requires a lot of written documentation in the form of the HACCP plan and forms filled in with specific information. These records serve a number of purposes including documentation, history, and identfying trends.

In addition, if the worst happened, and your company was faced with a recall, the records would serve as a way to identify the scope of the recall. Records can also serve as evidence in the event of legal action against your company. So records are extremely important.

The following records should be part of an overall HACCP plan:

- HACCP team,

- Product description,

- Flow diagram,

- Hazard analysis,

- Steps in the process that are CCPs,

- Critical limits for each CCP,

- Procedures for handling deviations,

- Monitoring activities,

- Records of actual deviations and corrective actions taken, and

- Verification activities.

You may be involved in keeping some these records. If so, be sure they are accurate, legible, signed, dated, and filled-in at the required times.

Verification

After the HACCP plan has been implemented, the HACCP team must verify that the HACCP system is working the way it is expected to work according to the written plan. This is called verification. The HACCP team uses various methods, procedures, or tests to validate that the HACCP system is operating as intended. These activities are not related to monitoring CCPs.

Verification may include direct observations of monitoring procedures, calibration of equipment such as thermometers, microbiological sampling of product, and reviewing records. HACCP systems are rarely perfect, and verification allows an opportunity to identify the flaws in the process and fine-tune them to correct the system.

In conclusion

As you can see by the information in this chapter, HACCP is a somewhat complicated process. When systematically broken down into many steps, however, it is a practical and manageable process. Food producers across the United States (and in other parts of the world) use the HACCP system to ensure the safe management and production of food, so consumers of the product, people just like you and me, can eat safe food every day.

NOTES

NOTES

HACCP REVIEW

1. HACCP stands for:
   a. Hazard Analysis and Critical Control Points
   b. Hazards and Allergens Cause Certain Problems
   c. Having any Affects Causes Control Policies
   d. None of the above

2. The purpose of HACCP is to:
   a. Make you do more work
   b. Eliminate potential hazards from food to make it safe to eat
   c. Clean equipment properly
   d. None of the above

3. HACCP identifies which types of hazards:
   a. Microbiological
   b. Chemical
   c. Physical
   d. All of the above

4. Which U.S. government agencies require HACCP:
   a. United States Department of Agriculture
   b. Food and Drug Administration
   c. Both (a) and (b)
   d. Neither (a) nor (b)

5. How many HACCP principles are there:
   a. Five
   b. Three
   c. Seven
   d. Eight

6. Which is not a principle of HACCP:
   a. Hazard analysis
   b. HACCP team
   c. Monitoring
   d. Recordkeeping

7. Which is an example of a physical hazard:
    a. Glass
    b. Cleaning chemicals
    c. Bacteria
    d. None of the above

8. Which type of hazard is the cause of foodborne illness:
    a. Physical
    b. Microbiological
    c. Chemical
    d. None of the above

9. A critical control point (CCP) is:
    a. Identifying hazards and preventive measures
    b. A point, step, or procedure in a food process at which a hazard can be controlled
    c. Product sampling and testing
    d. All of the above

10. HACCP records must be:
    a. Accurate
    b. Signed
    c. Dated
    d. All of the above

PERSONAL HYGIENE PRACTICES

As a person who works with and around food, you need to maintain a high standard of personal hygiene. Personal hygiene involves keeping yourself and your clothes clean, following specific cleansing procedures, wearing certain types of personal protective equipment (PPE), and following other sanitary work practices before and while working with food. Food workers are responsible for producing safe food, and good personal hygiene is one of many things needed to keep food safe from bacterial contamination.

After bacteria have been "killed" in a production process such as cooking, they can be reintroduced to food through contact with food workers' hands or clothing. Because this can happen, your need to maintain good personal hygiene is important to keep food safe. Good personal hygiene practices are regulated by the federal government under the Food and Drug Administration's (FDA) Current Good Manufacturing Practices, commonly called CGMPs.

CGMPs specify that all persons working in direct contact with food, surfaces that food might contact, and food-packaging materials must conform to hygienic practices to protect against contamination of food. By keeping yourself clean and following strict cleanliness rules at work, you keep the food you work with free from potential contamination with bacteria such as *Salmonella, Listeria,* and *E. coli.* These potentially dangerous food contaminants are described in more detail in the *Basic Microbiology & Foodborne Illness* chapter.

This chapter focuses on the personal hygiene habits and procedures you should follow every day to ensure your own safety, the safety of the food you help produce, and the safety of the people who eat it. Your personal hygiene can make the difference between a safe food product and one that can cause illness or even death to those who eat it.

The FDA's CGMPs outline essential elements of a good personal hygiene program. Each of the following CGMP topics will be covered in more detail in this chapter:

- Not working when sick or infectious;

- Maintaining adequate personal cleanliness;

- Wearing garments that protect food from contamination;

- Washing hands thoroughly when necessary;

- Removing all jewelry and other objects that might fall into food;

- Storing clothes and other personal belongings in designated areas separate from food production or utensil/equipment washing areas; and

- Keeping chewing gum, candy, beverages, food, and tobacco out of food production and storage areas.

Your company will provide specific guidance and rules on some of these practices, but you will have to take steps on your own to ensure that others are followed. However, you must take personal responsibility for following all of them. Failing to follow these good personal hygiene practices could endanger you and those who eat the food you make.

Not working when sick or infectious

The first rule you must always follow is never to go to work if you are contagious with bacteria that could be transmitted through food. This could cause sickness to others as a foodborne illness. The common cold cannot be passed on through food, but if you have symptoms such as diarrhea, cramps, or vomiting, you should stay home.

Some common bacteria that can be transmitted through food from people with these symptoms and cause illness include:

- *Salmonella,*

- *Staphylococcus,* and

- *E. coli*

See the *Basic Microbiology & Foodborne Illness* chapter of this handbook for more information on these bacteria and how to prevent their transmission.

Other potentially infectious hazards are open cuts and sores. Bacteria such as *Staphylococcus* can thrive in such wounds. These bacteria can then be passed on to food and cause serious illness in people who eat that food. Treat all cuts and sores as if they might be infected. Cover them with a protective covering such as a band-aid, with a rubber or latex glove over the top.

Maintaining adequate personal cleanliness

Personal cleanliness is something everybody should try to maintain, but it is absolutely essential for those employed in the food industry. The food you work with could become contaminated with bacteria or dirt if you are not clean.

Personal cleanliness begins at home. You are the only one who can ensure you maintain adequate cleanliness. Several simple habits you should follow to maintain adequate personal cleanliness include:

- Showering or bathing every day before you go to work. Bacteria and dirt can easily be found on unclean hair and skin;

- Putting on clean, laundered clothes after bathing and before going to work including socks, underwear, pants, and shirt. Soiled clothing can contaminate food;

- Using soap and water to wash hands thoroughly every time you use the restroom. Fecal material can become lodged under and around your fingernails;

- Trimming, cleaning, and filing fingernails frequently. The rough edges of cracked, chipped, or broken fingernails can contain dirt and bacteria; and

- Removing nail polish, fake nails, and jewelry before starting work. These items can harbor harmful contaminants.

Wearing garments to protect food from contamination

Your company will designate the kinds of clothes you must wear while working. The company may provide you with a uniform, smock, boots, apron, or other garments suitable for your job. Wear uniforms only at work to help prevent contamination. Never wear them outside the workplace.

Garments your company specifies must be worn when you are working, and in a manner that protects against the contamination of food, surfaces that food might contact, or food-packaging materials. This means you cannot modify the fit or style of the garment in a way that might expose your skin or other potential contamination in food work areas. For example, you cannot cut off the sleeves to expose more of your arms because hair could contaminate the food. In addition, when a garment becomes soiled or torn, it can no longer serve its purpose to protect against contamination. Therefore, you must change any protective garment when it becomes soiled or torn.

Keep smocks and other garments uncontaminated by removing them before using the restroom. Hang smocks, coats, or other outer garments in areas provided outside the entrance to the restroom. That way, you can go to the bathroom, wash your hands thoroughly, and put your garment back on without it becoming contaminated. If you wear it into the restroom and keep it on while using the facilities or hang it over the stall door, you will almost certainly contaminate it, making it unsafe to wear in food work areas.

Your company may also have sanitizing footbaths that you must walk through before entering a food work area. Never skip the footbath just because you are in a hurry or think your shoes or boots aren't dirty. Always use this and other wash facilities provided. If possible, leave your daily work footwear at your company in a locker or designated storage area. This helps keep potential contamination like dirt or manure outside the workplace and food work areas.

Finally, if you work with any type of allergenic food (like peanuts, almonds, soy sauce, or certain fish), be careful not to contaminate other food. Even minute traces of such allergenic foods, when transferred from your clothes or hands, could contaminate the next batch of food and cause illness or even death to an allergic person who eats it. Change your gloves, apron, smock, or other garments before beginning work on a food that does not contain that allergen. See the *Cross Contamination* chapter for more information on food allergens and preventing contamination.

Washing hands thoroughly when necessary

Washing our hands seems to be a simple procedure we often take for granted. When you work in the food industry, however, you must take handwashing more seriously to make sure you do a proper and thorough job each and every time you are supposed to. If not, disease-causing pathogens such as *Salmonella, Shigella, Listeria, E. coli* or hepatitis A virus can be spread from your hands to the food you work with.

Your company will provide handwashing facilities in several spots throughout the building. You must use these facilities to wash your hands at any of the following times:

- Immediately before starting at your workstation prior to putting on gloves, plastic sleeves, armguards, aprons, or other PPE.

- After coughing or sneezing into your hand(s).

- After touching or scratching any part of your skin, hair, eyes, or mouth.

- After adjustments are made to clothing such as coats, hard hats, hairnets, earplugs, shoes, etc.

- After using a tissue or handkerchief to wipe or blow your nose.

- After eating, drinking, or smoking.

- After using the restroom for any reason.

- After picking an item up off the floor.

- Before and after handling raw meats, poultry, or other raw foods.

- After handling items such as boxes, labels, garbage, brooms, hoses, etc. before returning to work on food production lines or handling racks of product.

- After each absence from your workstation.

The way you wash your hands is just as important as when you wash them. If you don't get them completely clean, you still may contaminate food. Pay special attention and wash your hands extremely thoroughly. Your handwashing technique should follow these simple steps:

1. Turn water on to the warmest temperature you can tolerate.

2. Take an adequate amount of germicidal/anti-bacterial soap or sanitizer your company provides at all handwashing stations.

3. Scrub vigorously, making sure that the soapsuds cover and clean every part of your hands.

4. Make sure you clean the webbing between your fingers, where dirt and germs can hide in the folds.

5. Make sure you clean under and around your fingernails, scraping any dirt out from under them, using a nail brush if one is available.

6. Rinse thoroughly, using the same warm water, making sure that all traces of soap are completely rinsed from your hands.

7. Use the disposable, single-use paper towel your company provides to dry your hands thoroughly. (Throw this away in the trash receptacle provided.)

In addition, when you wash your hands after using the restroom, two practices are strongly recommended. First, be especially certain to clean thoroughly under and around your fingernails. Fecal material can hide there even when you are careful, and could eventually contaminate the food you work with.

Second, use the "double-wash" method. This simply means you wash your hands twice, following the seven steps described previously. This method has been shown to reduce bacteria better than a single handwash after using the restroom.

Some handwashing stations may also be set up with hand dips, which contain sanitizers. If your work area has a hand dip station, use it frequently throughout the day. The solution should be at a concentration that will not contaminate food, so you do not need to rinse your hands after using the dip.

Removing all jewelry & other objects

You must remove all jewelry or other objects (such as watches, earrings, necklaces, bracelets, rings, and pens from your pocket) that might fall into food, equipment, or containers while you work. Jewelry can harbor skin, debris, hair, oil, and bacteria  that could cause foodborne illness. That's because jewelry rests next to skin and hair when you wear it, collecting potential contaminants in the grooves and crevices. Other personal items may also harbor dirt in cracks or crevices. Also, jewelry can easily break and fall into food. An earring may come apart or the face of a watch could pop off. These pieces could harm a consumer if eaten.

Before starting your work shift, remove every piece of jewelry you are able to. Leave it at home or store it in your locker. Sometimes, you may not be able to remove some pieces of hand jewelry, like a ring that is so tight it cannot come off your finger. If this is the case, your company may require the jewelry be covered by a rubber glove or other material. The covering must remain intact, clean, and sanitary to effectively protect against contamination of the food, food-contact surfaces, or food-packaging materials you work with.

Medic alert jewelry can stay on at all times. In a food work environment, the preferred variety of medic alert jewelry is a chain with a medic alert pendant, worn under the shirt.

Wearing & maintaining gloves & hair restraints

Gloves

Some jobs require the use of gloves for handling food. If your job requires gloves, the gloves you wear must be maintained in an intact, clean, and sanitary condition and made of an impermeable material (that's a material things can't soak or come through, like latex or plastic). If you wear gloves for handling food, you must use hand dip stations if they are provided and required. In addition, replace gloves with a new pair when they become torn or cut.

Never put gloves in your pockets or take them into washrooms, lunchrooms, or other unclean areas. Leave them at your workstation when leaving the work area, or get a new pair when you return.

Hair Restraints

Hairnets, beardnets, headbands, or caps are almost always required for employees working with food. Your company will provide these or require you to get a particular type. Either way, it is your responsibility to wear a hair restraint effectively, making sure it covers your hair completely, so none can escape and fall into food. People with long hair will find it easier to keep their hair contained if they first put it into a ponytail or braid, from which strands are less likely to come loose.

For employees who have beards, restraints should be worn for the same reasons as a traditional hair restraint. Special beard covers are available to restrain beard hair, and must be worn along with traditional hairnets or other hair coverings.

You should never perform any personal grooming, like combing your hair or smoothing your beard, in food production areas. Try not to touch your hair until after your work shift is complete, or you're in a locker room, changing area, restroom, or break area. If you do have to touch any hair, because of an itch or by accident, go to a restroom or changing area and:

- make sure your hair is completely under its hair restraint,

- wash your hands before returning to the work area, and

- get a new pair of gloves if you were wearing them.

Storing clothes & personal belongings out of food production and storage areas

At work, your company probably provides locker rooms, break areas, or similar storage areas where you must store non-work clothing and other personal belongings. These areas must always be separate from areas where food is exposed or where equipment or utensils are washed. Make sure you are aware of which areas are designated for this personal use and storage.

Belongings that should be kept in these designated areas include your:

- purse or wallet;

- jewelry;

- street clothes;

- cigarettes;
- magazines, newspapers, or books; and
- other non-food items.

You should never bring or use these items in areas where food may be exposed or where utensils or equipment are washed. In addition, you should never bring your lunch, soda, snacks, or other personal food items into these designated clothing and personal belonging storage areas.

Keeping chewing gum, candy, beverages, food, & tobacco separate from food or equipment areas

Chewing gum, candy, beverages, food, and tobacco must also be stored in designated areas. You cannot chew gum, eat candy, drink soda or other beverages, have snacks, or use tobacco products of any kind in your work area. You may not even bring those items into your work area because they could contaminate food.

Gum, candy, food, and drink cannot be stored with clothes and other personal belongings because they could attract bugs or vermin to clothing storage/changing room areas.

Contamination from these pests could be transferred from your clothes to food you make.

Your company will provide a separate designated storage area for these personal items to keep all food-related work areas safe from contamination. Your company probably has one or several refrigerated areas where employees can store their lunches, snacks, and beverages. Make sure you are aware of which areas of the company are designated for this use and storage.

Many companies have vending machines in lunchrooms or breakrooms. Eat, drink, and smoke or chew tobacco only in these designated areas as well. The proper storage and use of these snack items is strictly enforced in food companies to prevent food from becoming contaminated.

In conclusion

As you can see, there are many personal hygiene rules you must follow when working in food preparation or packaging areas. These rules are necessary to keep the food made by your company safe. You must follow these rules to ensure that safety.

NOTES

NOTES

Name _____

Date _____

PERSONAL HYGIENE REVIEW

1. Personal hygiene is:
   a. Keeping yourself clean
   b. Sanitizing of equipment and food work areas
   c. Coming to work on time
   d. None of the above

2. Which is not a good personal hygiene practice:
   a. Bathing daily
   b. Wearing hair restraints
   c. Wearing clean jewelry
   d. Handwashing

3. Good personal hygiene practices include:
   a. Not working when sick or infectious
   b. Wearing clean clothes to work
   c. Trimming, cleaning, and filing fingernails frequently
   d. All of the above

4. Which item can be brought into food work areas:
   a. Chewing gum
   b. Hair restraints
   c. Lunches
   d. Candy

5. Personal protective equipment (PPE) includes:
   a. Gloves
   b. Plastic sleeves
   c. Smocks
   d. All of the above

6. When wearing personal protective equipment (PPE), you should:
   a. Modify the cut or style
   b. Keep it clean and uncontaminated, and change it when it becomes torn or soiled
   c. Keep it on until the end of your shift, even if it's contaminated
   d. None of the above

7. How many steps are involved in a proper handwashing procedure:
   a. Seven
   b. Five
   c. One
   d. Two

8. You must always wash your hands after:
   a. Using the restroom for any reason
   b. Touching or scratching any part of your skin, hair, eyes, or mouth
   c. Eating, drinking, or smoking
   d. All of the above

9. Proper handwashing technique should include which of the following:
   a. Don't worry about cleaning under your fingernails or between fingers
   b. Using the coldest water possible
   c. Using the germicidal/anti-bacterial soap or sanitizer provided
   d. Drying your hands on your uniform

10. Where can lunches, drinks, candy, and chewing gum be allowed in food facilities:
    a. In food processing areas
    b. Only in lunchrooms, break areas, or other designated storage areas
    c. In clothing storage or changing room areas
    d. All of the above

PEST CONTROL

Pest control is a common sense sanitation issue. We all know we don't want pests such as birds, rodents, cockroaches, flies, or other animals and insects around food. We understand that they pose a food contamination risk that only a proper pest control program can control.

The pest control program is part of the company's overall sanitation plan, helping to prevent food from becoming contaminated. Pests can carry and transmit disease and filth. Consumers won't want to eat a company's food product if they suspect pests might have contaminated it.

Pest control is an issue which you, as an employee, have some control over. The level of your control may be even higher than you realize because most pest control issues are ones you work with or are around every day. Your company probably employs a professional pest control service to apply pesticides, set traps, and supervise the pest control program. However, employees like you are still responsible for carrying out many activities related to pest control. You need to be familiar with the pest control issues covered in this chapter, including:

- CGMPs and HACCP requirements,

- Common methods of pest control,

- Employee responsibilities, and

- Recommended daily work practices.

CGMPs and HACCP requirements

In the Food & Drug Administration's (FDA) Current Good Manufacturing Practices (CGMPs), the issue of pest control is addressed. In addition, pest control is considered a prerequisite program for Hazard Analysis and Critical Control Points (HACCP). That means any company operating a HACCP system is expected to have an effective pest control program in place.

The CGMPs specify some specific pest control rules. This includes keeping grounds clear of pests and inspecting and keeping outdoor food storage tanks free of pests. In addition, the CGMPs clearly state that "No pests shall be allowed in any area of a food plant."

That's why your company has a pest control program in place, and why you work within its guidelines to keep your workplace free of pests. The pest control program at most companies consists of a combination of efforts, such as those described in the next section.

Common methods of pest control

There is a saying that the best offense is a good defense. That statement holds true when it comes to pest control. The best way to combat pests is to prevent them from getting into the food production facility in the first place. To keep pests from getting in, the building should have no unprotected openings to the outside. Attractions like food and water around and in the facility should be eliminated as much as possible to prevent pests from finding a place of refuge or shelter.

However, despite the many precautions taken by your company and the professional pest control service, pests sometimes manage to get into the building. When they do, you need to know how your company is combatting them and how you can help in this effort.

There are several common methods of pest control that are used. These include different kinds of insect, rodent, and bird control measures. Some of these pest control measures are covered in this section.

Insect control

An extremely common method of insect control is the use of pesticides. Many pesticides are made of potentially harmful chemicals, so they must be handled, applied, and stored carefully. That's why many food companies employ professional pest control services. If a professional service is not used, the person who applies pesticides at your company must be thoroughly trained in pesticide application and hazards.

Because of the potential danger of pesticides, the CGMPs state that pesticide use is permitted only under precautions and restrictions that protect against the contamination of food, food-contact services, and food packaging materials. All food or food ingredients must be put away before a scheduled application of pesticides. You may be required to do this.

There are three basic methods of pesticide application which you should be aware of in case it affects you or your work area:

- Fogging to knock down adult insects,

- Fumigating, and

- Crack-and-crevice spraying, which is commonly used in lunchroom areas to control cockroaches.

Some of the controls you may be involved in to help reduce the hazards associated with pesticides as a result of handling, use, or storage include:

- Cleaning and sanitizing in or around all food work areas thoroughly after pesticide use and before starting food production.

- Storing pesticides in a safe place away from food areas in the proper manner for each pesticide.

- Wearing proper personal protective equipment (PPE) to protect yourself if you are the person trained to apply pesticides. This may include rubber gloves, mask or respirator, boots, or other PPE depending on the pesticide used.

A measure to prevent insects from entering your facility in the first place is to check incoming raw materials for pests and reject contaminated shipments from entering the plant.

Other methods of insect control include the use of non-chemical means, such as heat sterilization or placing insect electrocutors near entrances to the building. If you help maintain insect electrocutors, be sure to keep the bulb at the correct frequency, so the lights are effective at attracting insects. Empty the trays regularly as well.

Rodent control

Two main methods of controlling rodents in food production facilities are trapping and baiting. Trapping is using mouse, rat, or squirrel traps to capture and hold the rodent. Baiting uses traps with bait or areas with poisoned bait (rodenticides) that the rodent eats and carries away with it. Rodenticides poison the rodent and cause it to infect the nest and die. Baits are generally restricted to outside use.

Traps generally have several practical advantages over baits, including the fact that they:

- Provide a definite record of the results of rodents trapped as to what kind, how many, etc.;

- Identify problem areas in or around the facility;

- Pose no toxicity dangers like baiting does because the use of rodenticides; and

- Help determine the direction of travel and travel paths of rodents, thus making their elimination easier.

If you work around rodent traps, follow your company's standard operating procedures regarding rodent control. In addition, some general rules you can follow to help control rodents through trapping (if they don't conflict with your company's procedures) include:

- Check traps daily when mice are regularly being caught, less frequently (but not less than weekly) when mice are not regularly being caught.

- Reset accidentally sprung traps.

- Inform a supervisor or pest control person of lost or damaged traps, so they can be replaced.

- Place traps on solid surfaces that will not wobble when the rodent steps onto it. Traps on overhead surfaces should be tied or nailed down. Usually traps are placed next to walls, because rodents will usually travel along walls rather than going out into open areas.

- Never place traps near food production, processing, or packaging areas to prevent contamination.

- Remove dead rodents carefully by wearing gloves and disposing of the bodies in designated disposal containers.

You are probably not responsible for baiting, but if you are, follow the same guidelines for handling, use, and storage of rodenticides that have been outlined previously for pesticides. They also contain dangerous chemicals. As with pesticides, rodenticide use is permitted only under precautions and restrictions that protect against the contamination of food, food-contact surfaces, and food packaging materials.

If you work with or around rodent baiting, follow your company's standard operating procedures regarding the use of baits in rodent control. In addition, some general rules you can follow to help control rodents through baiting include:

- Use them only in designated and authorized areas (usually they are only used outside and not in food production areas),

- Inspect and change bait frequently, and

- Dispose of, or notify the proper person to dispose of, any rodent bodies you discover.

Bird control

A less common type of pest is the bird. Sparrows, starlings, and pigeons are all considered potential pest birds around food facilities. After eliminating potential food, water, and nesting areas, the best way to keep birds away is to follow common sense.

Common sense measures, such as those outlined in the next section, *Recommended daily work practices*, include measures that anyone who works in a food production or storage facility should follow every day to support pest control at their company.

Recommended daily work practices

Perhaps the best method in a good pest control program is to prevent pests from entering the food production facility to begin with. Many of these recommended daily work practices are ways that you, as an employee, can help keep your company free of pests.

These recommended work practices should be followed every day to support the pest control program in effect at your company.

1. Always maintain good housekeeping in all areas of the facility. Keep all work and storage areas clean, sanitary, and dry. Promptly clean up any food ingredient spills that could attract pests.

2. Keep garbage areas clean both inside and outside the building. Make sure any leaks from garbage compactors are cleaned up.

3. Inform pest control personnel or your supervisor whenever you see evidence of an infestation of any kind.

4. Follow good personal hygiene practices by keeping food, snacks, beverages, and other belongings in a locker or designated storage area.

5. Never prop outside doors open for ventilation. Always keep doors and unscreened windows closed unless immediately in use, or unless they have an air curtain that can prevent pests from entering.

6. Closely inspect incoming raw materials. Reject those materials that harbor pests.

7. Frequently check devices designed to catch pests. Empty, clean, or replace them on a regular basis to maintain their effectiveness, or notify somebody who can.

8. Storage areas are especially attractive to pests and require special attention. Use good storage practices by rotating stock, storing items away from walls, and cleaning up any spills immediately.

Follow these general rules and the pest control guidelines outlined in this chapter. They should help keep your work environment free of pests and the products you make safe.

NOTES

NOTES

Name _____

Date _____

PEST CONTROL REVIEW

1. Pests include which of the following:
    a. Insects
    b. Birds
    c. Rodents
    d. All of the above

2. Pest control is part of your company's:
    a. Sanitation program
    b. Equipment maintenance
    c. Microbiological testing
    d. None of the above

3. Pests are attracted to:
    a. Food
    b. Water
    c. Garbage
    d. All of the above

4. The best way to combat pests is to:
    a. Combat them once they're inside the facility
    b. Keep them from getting into the food facility in the first place
    c. Detect them once they are in the food product
    d. None of the above

5. To keep pests from getting in:
    a. The facility should be tightly constructed
    b. Eliminate food and water sources
    c. Keep doors and windows shut
    d. All of the above

6. Insect control methods include:
    a. Insecticides
    b. Insect electrocutors ("zappers")
    c. Heat sterilization
    d. All of the above

7. The methods used to apply pesticides include:
   a. Fogging
   b. Fumigating
   c. Crack-and-crevice spraying
   d. All of the above

8. The main method of rodent control is:
   a. Baiting
   b. Trapping
   c. Both (a) and (b)
   d. Neither (a) nor (b)

9. A general rule you can follow to help control rodents through baiting include:
   a. Using them only in designated and authorized areas
   b. Inspecting and changing bait frequently
   c. Disposing of, or notifying the proper person to dispose of, any rodent bodies you discover
   d. All of the above

10. Which is not an example of good pest control:
    a. Cleaning up spills
    b. Leaving doors and windows open
    c. Cleaning up garbage
    d. Using pesticides

PEST CONTROL

SANITATION

The topic of sanitation does not only include cleaning equipment and work areas. Sanitation refers to *all* of the practices and procedures used to keep a food production facility clean and the food produced free of contaminants and potentially harmful bacteria.

Proper sanitation is necessary to keep the food your company produces safe for human consumption. Sanitary food production not only makes good business sense, but it is also regulated by several different United States government agencies and standards.

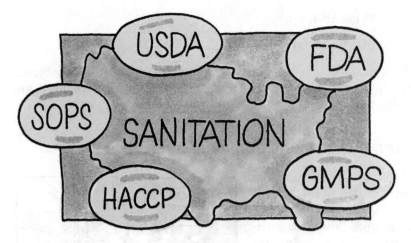

The Food and Drug Administration's (FDA) Current Good Manufacturing Practices (CGMPs) outline many sanitation procedures and practices. In addition, the United States Department of Agriculture (USDA) requires some food manufacturers and processors to implement Standard Operating Procedures (SOPs) for sanitation. In fact, sanitation SOPs are considered an important part of a Hazard Analysis and Critical Control Points (HACCP) system. HACCP is a food safety system covered in its own chapter in this handbook.

As you can see, sanitation is considered an essential operation in all food production facilities, including the one where you work. Sanitation is necessary to ensure the production of food free of contaminants and microorganisms. Some of the topics related to sanitation that probably effect your job include:

- Personal hygiene and cleanliness,
- Equipment and work area cleaning,
- Housekeeping,
- Use and storage of cleaning chemicals and equipment,
- Work procedures and controls related to sanitation,
- Pest control, and
- Sanitation SOPs.

This chapter covers each of these sanitation-related topics. Understanding your role and responsibilities will help you keep your company sanitary and the food safe.

Personal hygiene and cleanliness

The most important part of the CGMPs that affects you is personal hygiene and cleanliness. This section describes expected cleanliness practices all food company employees must follow to keep themselves and their food workplace clean and sanitary.

Personal hygiene practices include activities like washing your hands, wearing clean clothes, bathing, and trimming your fingernails. They also include work-specific activities like wearing a hairnet, using footbaths and hand dip stations as required, and washing your hands every time you return to your work area.

Personal hygiene practices are so important to a clean and sanitary food work environment that they are covered in more detail in a separate chapter in this handbook. See the *Personal Hygiene* chapter for more information on important personal hygiene practices.

Equipment and work area cleaning

One of the most common images associated with sanitation is cleaning and sanitizing equipment and food-contact surfaces. Cleaning removes visible soil and food residue, while sanitizing kills bacteria with a chemical. The reason for cleaning and sanitizing is to prevent the growth of potentially harmful bacteria on equipment and utensils.

The food residue left on equipment is a source of food for bacteria, which can grow very rapidly. If equipment is not cleaned properly, the bacteria will contaminate the next batch of food that comes into contact with the unclean equipment. The sanitizer used to kill the bacteria can't do its job properly if food debris is left on the equipment. This prevents the sanitizer from reaching all of the bacteria. See the *Basic Microbiology & Foodborne Illness* chapter for more information on the bacteria whose growth can be prevented through proper cleaning and sanitizing.

Most food company employees are involved in cleaning and sanitizing activities. You should perform them regularly throughout the day to maintain sanitary conditions to ensure the safe production of food. For example, if you work on a food production line, you may be the person responsible for cleaning it during or after your work shift. Keep your work area clean and tidy at all times by picking up garbage, cleaning up spills right away, and putting items in their designated storage areas to decrease the chance of contamination.

Your company might have a daily sanitation or cleaning crew, but keeping work areas clean throughout the day is still very important. After cleaning, you must sanitize the food-contact work surface with special sanitizing chemicals.

You may also be responsible for taking pieces of food processing equipment apart and cleaning and sanitizing them according to specific procedures. Cleaning and sanitizing involves a basic three-step process:

1. A rough clean to wipe all visible residue off with a cloth and/or brush;

2. Cleaning with a chemical, like a detergent; and

3. Sanitizing with a chemical, like a bleach solution.

You may or may not have to rinse after the second and third steps, depending on which chemicals are used and at what strength. The chemical strength and contact time (the amount of time the cleaner or sanitizer is in contact with the surface being cleaned or sanitized) are very important.

You must follow established contact times to ensure the chemical accomplishes its cleaning or sanitizing job thoroughly. Don't rush through it because you think you don't have enough time or it looks clean to you. Remember you can't see bacteria without using a microscope. You must make enough time to keep the food your company produces safe from contamination with bacteria.

Cleaning and sanitizing chemicals can be applied in different ways. They can be:

- Mixed with water for cleaning by hand with a brush or cloth,

- Metered into a water system that can be used through a hose,

- Applied as a foam, or

- Used in a Clean In Place (CIP) system where cleaning and sanitizing chemicals are circulated through an enclosed system to cleanse tanks, lines, and other parts of the equipment.

Follow the unique steps and practices for each type of cleaning and sanitizing procedure you have been trained to perform.

Some specific rules that relate to cleaning and sanitizing activities are found in FDA's CGMPs. Your company has developed procedures and practices for cleaning and sanitizing activities according to these rules. These cleaning procedures are specific to your facility and the food being produced.

If you are expected to perform any of these cleaning activities, you should receive instructions and training on the proper procedures for your workplace or work area.

Your duty is to follow the steps and procedures you are trained to use. Always use the types of cleaning equipment and chemicals specified for each job. Never skip a step in the process or use a substitute cleaning or sanitizing chemical unless approved by your supervisor.

Use and storage of cleaning chemicals and equipment

The use and storage of cleaning and sanitizing chemicals and equipment is another important sanitation activity. Because we need to clean and sanitize equipment and work surfaces, we also need the proper chemicals and equipment to do the job.

Cleaning equipment and chemicals are necessary to keep food companies clean. However, cleaning equipment and chemicals can be a contamination hazard if not used, handled, and stored properly. For example, bleach can create a toxic contamination hazard if left in or stored too close to food production or processing areas. If bleach accidentally got into a batch of food, it would pose a serious health hazard that could make a person very ill.

Many other cleaning and sanitizing chemicals are also toxic and dangerous to people if accidentally ingested. The cleaning equipment and chemicals used for sanitizing must always be used according to label directions and your company's sanitation procedures.

Cleaning equipment like mops, brushes, and rags must also be cleaned before being put away. Dirt or food left on cleaning equipment will provide a source of food for bacteria to grow. The next time you use that mop, brush, or rag you will actually be spreading bacteria onto manufacturing equipment, rather than removing it, as cleaning and sanitizing are intended to do. Therefore, all types of cleaning equipment should be cleaned and hung to dry after each use to prevent the growth of bacteria which thrive in damp and wet conditions. Some companies use disposable cloths or other equipment to avoid possible contamination during cleaning and sanitizing.

Cleaning and sanitizing chemicals must also be stored safely away from areas where food is made. The proper use and storage of these chemicals is so important that it is regulated under FDA's CGMPs.

Your company has designated procedures for use and storage of cleaners, sanitizers, and equipment. Always use cleaning equipment and chemicals according to directions and store them properly when cleaning is complete.

Some companies have cleaning equipment just for certain areas of the facility to prevent cross contamination of work areas. For example, perhaps certain buckets, mops, and brushes are used only in raw food areas, while a different set is used only in finished product areas. If this is the case, the equipment may be color-coded or labeled accordingly. Only use the proper set of equipment for that work area you're cleaning and store it in its designated place.

Also, don't forget to use proper personal protective equipment (PPE) when cleaning and sanitizing. Wear gloves, eye protection, or other PPE that may necessary while using the chemicals you work with. Read and follow label instructions carefully.

The *Cross Contamination* chapter in this handbook has a section with more information on the proper storage and use of cleaning chemicals. See that chapter for more detailed information on this sanitation topic.

Work procedures and controls related to sanitation

Work procedures and controls are an essential part of sanitation, because the potential for the growth of bacteria or other contamination of food is high.

Work procedures and controls cover all of the processes at your company designed to maintain sanitary conditions during receiving, inspecting, transporting, storing, preparing, manufacturing, and packaging food.

The overall sanitation at your company is under the supervision of someone assigned to be responsible for this function. They determine the nature and content of the work procedures and controls needed to produce safe food based on regulatory requirements and food safety principles. You and your fellow employees will implement many of the procedures. You are the ones likely to be inspecting incoming food for pests or other contaminants, making sure food is kept refrigerated at specific temperatures, or performing general housekeeping tasks.

For example, all raw materials and other ingredients arriving at your company must be inspected to ensure they are clean and suitable for use in food and stored under conditions that protect against contamination and deterioration. While you most likely are not the one who determines the method of inspection or storage, you may be the person who performs the inspection or puts the ingredient into storage.

Whoever inspects incoming material or puts it into storage must ensure sanitary conditions are maintained. That person must follow your company's sanitation rules and procedures for these processes to meet the specific sanitary requirements. You will be trained in any processes such as these.

Other common processes and controls regulated in the processes and controls section of the CGMPs include:

- Washing raw materials;

- Monitoring raw materials for bacteria, natural toxins (like aflatoxin), and pests;

- Storing ingredients in a way that protects against contamination;

- Cleaning equipment;

- Producing and storing food in ways that minimize the growth of bacteria or the contamination of food; and

- Using time and temperature controls to keep food safe.

Time and temperature controls are so important that a whole chapter has been devoted to the topic. See the *Time & Temperature Controls* chapter for more detailed information on how those may be used as sanitation processes and controls. HACCP is also important to sanitation. See the *Hazard Analysis & Critical Control Points* chapter for more information on HACCP.

Pest control

Pest control is a common sense sanitation issue. We all know we don't want pests such as birds, rodents, cockroaches, flies, or other animals and insects around food. We understand they pose a food contamination risk that only proper sanitation can control.

The CGMPs specify some pest control rules, such as keeping grounds clear of pests and inspecting and keeping outdoor food storage tanks free of pests. In addition, the CGMPs define guidelines to keep food companies free of pests at all times.

To help keep your company free of pests, follow your company's rules and those mandated by the CGMPs, such as the following:

- Keep personal food, snacks, and beverages out of food work areas and clothing storage areas,

- Never prop open doors or windows without screens in the plant,

- Use pesticides and rodenticides only according to established procedures, and store them in a designated area, and

- Notify a supervisor or clean up any spilled food or ingredients that can be an attraction to pests.

Because it is such an important sanitation issue, pest control is covered in a separate chapter of this handbook. See the *Pest Control* chapter for more detailed information on how you can help prevent pests in your facility.

Sanitation SOPs

Standard operating procedures (SOPs) are simply established procedures for conducting a process. SOPs are usually in writing and specify the method, duration, and frequency of the given process the SOP applies to. SOPs should cover the who, what, when, why, and how for each procedure. Every company must develop its own unique set of SOPs for given processes because SOPs are specific to the activity and the environment in which it is performed.

There are SOPs for many activities in food manufacturing. Sanitation SOPs are extremely common for most food companies, since they have so many sanitation issues to deal with. Some of the sanitation issues described previously in  this chapter may be an SOP at your company. Your company may have established other sanitation SOPs.

The USDA requires all federally inspected food facilities to develop, implement, and maintain a written sanitation SOP plan. The content of the plan is simply a written record of the actual sanitation processes and procedures that the company has in place and uses every day.

The written sanitation SOP plan, if it is required for your company, must include the following:

1. A description of each daily sanitation procedure conducted before, during, and after manufacturing food to prevent contamination of product.

2. Sanitation procedures conducted before food can be made each day. These procedures must address the cleaning of food-contact surfaces, equipment, and utensils.

3. How often each sanitation procedure will be done.

4. Identification of the individuals responsible for implementing and monitoring daily sanitation activities.

5. Records maintained on a daily basis to document the implementation and monitoring of sanitation SOPs and any corrective actions taken.

6. The sanitation SOP plan signed and dated by a person with overall authority at your company.

The sanitation procedures outlined in your company's written sanitation SOP plan must be monitored on a daily basis. You may be required to participate in the implementation or evaluation of your company's sanitation program. For example, you may clean, inspect, teardown, or setup food processing equipment. You may also have to document any deficiencies in sanitary conditions or sign your name on a recordkeeping form. If so, you will be trained to follow the specific requirements of your company's sanitation SOP program.

In conclusion

Sanitary food production clearly presents many challenges to any company. There are several ways to keep the food production environment clean and safe. As long as you follow instructions and are aware of the sanitation issues you work with, you can contribute to the safe production of food.

NOTES

NOTES

Name _____

Date _____

SANITATION REVIEW

1. Sanitation refers to:
   a. Only cleaning of equipment and food work areas
   b. Only sanitizing of equipment and food work areas
   c. All of the practices and procedures used to keep the facility clean and the food produced uncontaminated
   d. Cleaning and sanitizing of non-food work areas

2. Which of the topics are related to sanitation:
   a. Personal hygiene and cleanliness
   b. Equipment and work area clothing
   c. Pest control
   d. All of the above

3. Which does not require strict sanitation procedures be followed:
   a. Food production equipment
   b. Employees' automobiles
   c. Food production work areas
   d. Employees' personal hygiene

4. Food residue left on equipment is:
   a. A source of food for bacteria
   b. A source of chemical contamination
   c. Ok, it does not need to be cleaned off
   d. All of the above

5. How many steps are in a basic cleaning and sanitizing procedure:
   a. One
   b. Two
   c. Three
   d. Four

6. The basic steps of cleaning and sanitizing include:
   a. A rough clean
   b. Cleaning with a chemical
   c. Sanitizing
   d. All of the above

7. Methods of applying cleaning and sanitizing chemicals include:
   a. Mix with water and apply with a brush
   b. Meter and apply through a hose
   c. Apply as a foam
   d. All of the above

8. Cleaning and sanitizing equipment and chemicals should be:
   a. Left any old place in the plant after use
   b. Used, handled, and stored properly
   c. Put away when still wet
   d. None of the above

9. When storing cleaning and sanitizing equipment, make sure to:
   a. Clean the equipment after each use
   b. Dry the equipment thoroughly before putting it away
   c. Store the equipment only in its designated storage area
   d. All of the above

10. When using cleaning and sanitizing chemicals always:
    a. Read and following the label instructions carefully
    b. Do whatever way you want
    c. Use as much or as little as you want
    d. None of the above

TIME & TEMPERATURE CONTROLS

Time and temperature controls are food production processes set up through the use of time limits and temperature monitoring and maintenance within a certain range. Time and temperature controls are used in the food industry to ensure a food product is cooked according to a specific thermal process or stored at the correct temperature.

The purpose of time and temperature controls is to control the growth of potentially harmful bacteria. Time and temperature controls are used to destroy or prevent the growth of pathogens such as *Listeria*, *Salmonella*, and *E. coli* in food, as well as to slow spoilage.

There are many different types of harmful bacteria that can be controlled by time and temperature. See the *Basic Microbiology & Foodborne Illness* chapter for more information on the potentially harmful bacteria that time and temperature are meant to control.

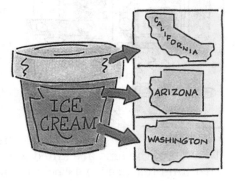

This chapter covers the following aspects of time and temperature controls:

- Common time and temperature controls;

- Relation to CGMPs, HACCP, and thermal processing; and

- Work practices to ensure time and temperature controls are maintained and complied with.

Common time and temperature controls

There are many different types of time and temperature controls and processes. Different controls and processes are used for different reasons on different types of food. Which heat process used in making a product depends on the type of food being processed or produced, what packaging is to be used, how it will be sold, and many other factors.

Several of the most common time and temperature processes are summarized here:

Blanching — A pre-packaging heat treatment of food for a time and at a temperature to partially or completely inactivate the naturally occurring enzymes and to effect other physical or biochemical changes in the food.

Cooking — Cooking is the most common time and temperature process that is used in food processing. Temperature is used to ensure that the food undergoes the correct heat process in cooking. Common cooking practices include baking, boiling, par-boiling, frying, steaming, broiling, roasting, and microwaving. Each method heats food to a temperature at which pathogens are killed and produces the characteristic flavor and texture we associate with cooked food.

Holding and storing — Holding time and temperature specifies the temperature to be reached and the time that temperature must be held as part of a heat process. For instance, in the canning process, you are required to hold the food at a certain temperature for a certain period of time. The holding time and temperature are determined by experts to ensure that the process kills harmful bacteria.

Freezing — Freezing involves cooling food to a temperature below 28 degrees Fahrenheit. While normal freezing is not considered a "kill" process, "quick freezing" has proven effective in killing some bacteria in food. Normal freezing merely stops the growth of bacteria, but does not kill it. That's why you must be careful when thawing frozen food, because the bacteria can begin to grow again. Freezing for extended periods of time (over two months) can kill some parasites, like *Trichina* found in pork. However, freezing alone cannot be relied upon to kill potentially harmful bacteria.

Irradiation — Exposing food to radiation for a period of time can kill various types of contaminating bacteria. This is not a temperature control, but it performs the same purpose of killing bacteria, so it is included here.

Pasteurization — This process involves heating food to a certain temperature, holding it at that temperature, then cooling it again. The time the product needs to be held will depend on the product and the temperature being used. The higher the temperature, the less time is needed to kill the bacteria. It can range from 15 seconds in some "high temperature short time" (HTST) processes used for milk, to 30 minutes for a vat batch pasteurization in a cheese plant. The time and temperature requirements are calculated by experts to ensure the process is adequate to kill harmful bacteria. Pasteurized product is not sterile and must still be stored under refrigeration, like milk.

Thermal processing — Although all heat processes are thermal processes, the term "thermal processing" is given to specific processes like canning and bottling. Control of this process is so important that there are special regulations for the thermal processing of low acid foods packed in hermetically sealed containers. This process involves filled containers being heated to a specific temperature for a specific length of time. The finished product is considered sterile and can be stored without refrigeration. Just a few examples of this type of food include canned stews, soups, and pouched juices.

Refrigerating — Refrigerating is cooling food to a temperature low enough to prevent the growth of potentially harmful bacteria. Most coolers are set around 40 degrees Fahrenheit. In some plants, the food processing rooms are also refrigerated to prevent bacterial growth while making food. It is very important that the temperature of coolers and refrigerated processing rooms is monitored to make sure the correct temperature is being maintained.

Ultra High Temperature (UHT) — UHT is a process in which food is heated in complicated equipment to a very high temperature for a very short time. The product is then packaged in sterile conditions. The finished product is considered shelf stable and does not need to be refrigerated.

All time and temperature controls must always be carried out properly according to established procedures to adequately protect food. Always make sure the times and temperatures specified are reached when working with foods having time and temperature controls.

Relation to CGMPS, HACCP, and thermal processing

Time and temperature controls are a significant part of the Food and Drug Administration's (FDA) Current Good Manufacturing Practices (CGMPs), the United States Department of Agriculture's (USDA) Hazard

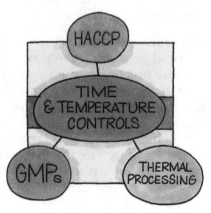

Analysis and Critical Control Points (HACCP), and the FDA's Thermally Processed Low-Acid Foods regulations. This section describes how time and temperature controls relate to each of these government regulations.

FDA's CGMPs

The FDA's CGMPs specify many production and process controls to prevent contamination of food. The CGMPs cover personal hygiene practices, pest control, sanitation, as well as specific time and temperature controls.

Some time and temperature controls specified include maintenance of proper temperatures during refrigeration, cooking, and freezing. Also mentioned are the use of sterilization, irradiation, pasteurization, and other methods to control the growth of bacteria.

HACCP

Similar to the CGMPs, time and temperature controls are also specified under HACCP as a method to eliminate or prevent unwanted growth of bacteria in food. Time and temperature controls frequently are Critical Control Points (CCPs) in a HACCP system. This is because time and temperature controls are often points in the production process where the contamination of food is prevented or eliminated and the safety of food is maintained.

Thermal processing

Thermal processing is a form of heating typically used in the processing of low-acid foods packaged in hermetically-sealed containers, such as cans and bottles.

A minimum thermal process is the application of heat to food, either before or after sealing in a hermetically-sealed container, for a period of time and at a temperature scientifically determined to be adequate to ensure destruction of microorganisms of public health significance.

The procedures for proper thermal processing are so complex that they have their own separate regulation, called *Thermally processed low-acid foods packaged in hermetically- sealed containers.* If you work at a company that uses thermal processing under this regulation, you will be trained in all of the correct procedures and practices related to thermal processing. As with all time and temperature controls, the most important thing is to make sure you always follow the established practices and procedures at your company.

Work practices to ensure time and temperature controls are maintained and complied with

As an employee, you are closest to the actual production process, and therefore have some control over the maintenance or monitoring of proper times and temperatures of products. By following certain work practices, you can make sure time and temperature controls are always maintained.

Follow these simple rules to ensure correct time and temperature controls are met for the safety of food products:

1. Always follow your company's established work procedures and practices for the food product you are working on.

2. Don't prop open doors to a refrigerator, freezer, or any other cooled work areas.

3. Never alter the specified time to heat, blanch, cook, or cool a product.

4. If in doubt as to whether a food has reached the correct processing temperature, use a thermometer to check.

5. When timing of procedures or practices is necessary, pay careful attention to the clock to be sure to meet the correct time requirement.

6. If you think a time or temperature control may have been violated, tell your supervisor or quality control person. He or she can test the batch or make a decision as to what to do.

7. If you are responsible for the time and temperature controls at a CCP under a HACCP system, you may be required to do several different tasks. This could include taking temperatures of various products, documenting temperatures on a recordkeeping form, monitoring a temperature chart, and signing and dating records.

Following these common sense time and temperature guidelines will help you keep the food your company produces safe and free of harmful bacterial contamination.

NOTES

NOTES

Name _____

Date _____

TIME & TEMPERATURE CONTROLS REVIEW

1. Time and temperature controls are food production processes set up through the use of:
   a. Time limits
   b. Temperature monitoring and maintenance within a certain range
   c. Both (a) and (b)
   d. Neither (a) nor (b)

2. Time and temperature controls are used to properly:
   a. Cook food
   b. Cool food
   c. Store food
   d. All of the above

3. Time and temperature controls include:
   a. Cooking
   b. Freezing
   c. Pasteurization
   d. All of the above

4. The purpose of time and temperature controls is to:
   a. Produce food faster
   b. Control the growth of potentially harmful bacteria
   c. Create a warm, moist environment
   d. Make food taste good

5. Which bacteria can be controlled by time and temperature:
   a. Listeria
   b. E. coli
   c. Salmonella
   d. All of the above

6. Which is the most commonly used time and temperature control in food processing:
   a. Cooking
   b. Irradiation
   c. Ultra High Temperature (UHT)
   d. All of the above

7. Cooling food to about 40 degrees Fahrenheit is called:
   a. Freezing
   b. Pasteurizing
   c. Refrigerating
   d. All of the above

8. The first rule to follow for time and temperature controls is:
   a. Being close to target is good enough
   b. Always follow your company's established procedures and practices for the food you are making
   c. If you make a small mistake on the time or temperature, it won't matter much
   d. None of the above

9. To make sure correct time and temperature controls are met:
   a. Don't prop open doors to a refrigerator, freezer, or any other cooled work areas
   b. Never alter the specified time to heat, blanch, cook, or cool a product
   c. Use a thermometer to check if a food has reached its correct processing temperature
   d. All of the above

10. If you think a time or temperature control may have been violated:
   a. Tell your supervisor or quality control person, so they can test the batch or make a decision as to what to do
   b. Throw the batch away when no one is looking
   c. Don't tell anyone
   d. None of the above

GLOSSARY OF FOOD SAFETY-RELATED TERMS

Anti-bacterial — Directed at or effective against bacteria. Soap that is anti-bacterial eliminates or kills bacteria.

Aseptic — Sterile; free of pathogenic organisms.

Bacteria — Minute microorganisms containing one cell each that reproduce by cell division. There are several types of potentially harmful bacteria associated with causing food-borne illness when the number of bacteria present in food reach a sufficient number. Some examples include *E. coli* and *Staphylococcus aureus*.

Biological hazard — Also known as a microbiological hazard. In food safety, this term is generally used to refer to the bacteria that pose a risk of causing foodborne illness if they reach a sufficient number on or in food contaminated with them.

Calibration — The adjustment of a testing machine or equipment to achieve accurate measurement.

Chemical hazard — A hazard posed to food by chemicals in the food production or processing workplace.

Communicable disease — A disease that can be passed on to another person.

Contagious — The condition of a disease such that it can be passed on from one thing to another.

Contaminant — Anything that can get into food that is not supposed to be there. Types of contaminants include glass, metal, wood, bone, hair, jewelry, bacteria, etc.

Fecal — Anything of, relating to, or constituting feces.

Feces — Bodily waste discharged through the anus.

Food & Drug Administration (FDA) — A United States government regulatory agency that is part of the Public

Health Service, under the U.S. Department of Health & Human Services. This agency regulates all foods, except meat, poultry, and egg products.

Food-contact surfaces — Surfaces that food comes into contact with. This includes utensils and surfaces of equipment.

Food Safety & Inspection Service (FSIS) — The branch of the United States Department of Agriculture (USDA) that is in charge of regulating meat, poultry, and egg safety.

Footbath — A container of water and/or germicidal solution provided to dip and clean feet/boots in food work areas to ensure the maintenance of clean feet in food processing areas.

Gastroenteritis — Inflammation of the lining membrane of the stomach and the intestines.

Germinate — To sprout or develop.

Germicidal — Kills germs.

HACCP plan — A document written for each food product in a facility regulated under HACCP. It describes each food product, it's manufacturing process, and identifies all possible biological, chemical, and physical hazards associated with that product for each process step. It also identifies critical control points, who is responsible for the critical control point, how to monitor it, what to do if something goes wrong, and what records to keep.

Hand dip — A container of water and/or germicidal solution provided to dip and clean hands in food work areas to ensure the maintenance of clean hands while working with food.

Hair restraints — Anything used to pull back and/or capture hair to the head and protect it from escaping or falling into the food. Common types of hair restraints include hairnets, caps, beardnets, or other devices designed to restrain hair in the workplace.

Hermetically-sealed container — A container designed and intended to be tightly sealed against the entry of air, liquids, and microorganisms to maintain the sterility of its contents after processing.

Hygienic — Clean and sanitary.

Impermeable — Not able to be penetrated by air or water. Impermeable gloves are required for some food production processes to maintain sanitary conditions.

Low-acid foods — Any foods, other than alcoholic beverages, with a finished pH greater than 4.6 and a water activity (a_w) greater than 0.85. Tomatoes and tomato products having a finished pH less than 4.7 are not classed as low-acid foods.

Microbes — Bacteria, molds, spores, and other microorganisms.

Microbiology — The science of studying microbial life forms.

Microorganisms — Living things that are so small that they can only be seen with the help of a microscope.

Minimum thermal process — The application of heat to food, either before or after sealing in a hermetically-sealed container, for a period of time and at a temperature scientifically determined to be adequate to ensure destruction of microorganisms of public health significance.

Pathogenic — Causing illness or death of a living thing through the introduction of a disease-causing microorganism.

Personal protective equipment (PPE) — The items worn as the outer layer during work with food. Common types of PPE include lab coats, uniforms, gloves, aprons, plastic sleeves, and boots.

Pest — Any objectionable animals or insects including, but not limited to, birds, rodents, flies, and cockroaches.

Pesticide — A solution made of toxic chemicals used to kill pests such as birds, rodents, and insects.

Physical hazard — Any object that can get into food during the production process and contaminate the food product. Common types of physical hazards include metal, glass, wood, dirt, stones, hair, etc.

Rodenticide — A solution made of toxic chemicals used to kill rodents.

Sanitize — To adequately treat food-contact surfaces by a process that is effective in destroying or reducing the number of undesirable microorganisms without adversely affecting the product or its safety for the consumer.

Thermal processing — All heat processes are thermal processes, but the term "thermal processing" is given to specific processes like canning and bottling or sealing in other hermetically-sealed containers, like pouches. There are special regulations for the thermal processing of low-acid foods packed in hermetically- sealed containers.

Toxin — A poisonous substance that is a product of the metabolic activities of a living organism and is usually very unstable.

United States Department of Agriculture (USDA) — A United States government regulatory agency that regulates food safety for animal food products.

Vermin — Pests, especially rodents.